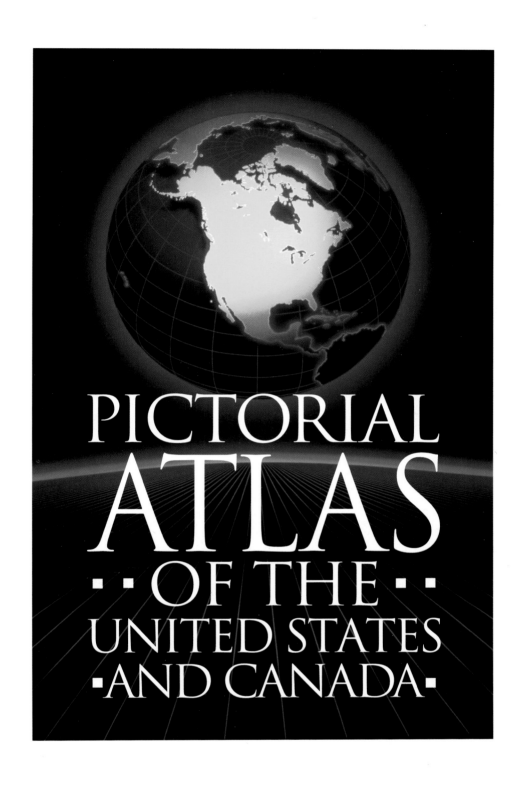

PICTORIAL ATLAS
·· OF THE ··
UNITED STATES
·AND CANADA·

BY KATHIE BILLINGSLEA SMITH

TRIDENT PRESS INTERNATIONAL

Trident Press International, Naples, Florida, 33940 U.S.A.
© Cover: Imtek Imagineering/Masterfile Corporation
ISBN 1-888777-07-9
Created and manufactured by Ottenheimer Publishers, Inc.
© 1996 Ottenheimer Publishers, Inc.
10 Church Lane, Baltimore, Maryland 21208
All rights reserved
Printed in Italy
AT005B

CONTENTS

Number in italics indicates map page number.

AN ATLAS is, by definition, a book of maps. The name "atlas" was first used by Gerhardus Mercator, a Flemish mapmaker of the 1500s. However, Claudius Ptolemy, an astronomer and geographer who lived in Egypt in the A.D. 100s, is credited with publishing the first atlas. His hand-drawn world map was remarkable for its time, but nonetheless exaggerated the space between Spain and China and underestimated the size of the ocean. Because of this mistake, Christopher Columbus was encouraged to make his historic voyage in 1492.

So, in a broader sense, an atlas is also a doorway to the world and to people and places unseen. An atlas educates us and inspires us to move beyond our own perimeters and explore the outside world. The *Pictorial Atlas of the United States and Canada* offers an overview of two of the world's most highly developed and economically productive countries. The United States and Canada are vast, varied lands of great beauty and tremendous natural resources. With this book you can travel from the treeless tundra of Auyuittuq National Park on Baffin Island in the Northwest Territories to the sparkling black sands of Kaimu Beach in Hawaii. You can scale the majestic peaks of the Canadian Rockies, sense the solitude of a Louisiana bayou, or experience the Big Sky Country of Montana. An atlas is a book of facts, but it is also a book of dreams.

This atlas is packed with information about America's fifty states and its territories, as well as Canada's ten provinces and two territories. A page is devoted to each state, province, and territory, featuring information about that area's history, major cities, special attractions, natural wonders, and geography. A summary of the state, provincial, and territorial nicknames, birds, flowers, trees, mottos, songs, flags and suggested scenic routes is also given. Full-color photographs provide a closer look at many of the area's attractions. For additional information about a particular area, a tourism bureau phone number is given on each page.

Colorful, detailed maps of each state, province, and territory are found in the latter half of this book, along with maps of major U.S. cities. In the back of the book is an index to help you find specific locations on the maps. An overall map of the United States and a chart showing mileage between major U.S. cities are included to assist you with travel planning.

This atlas also features numerous quick-reference charts. These tables list major products and industries, record dates of entry, and rank the states, provinces, and territories by area and population. Fascinating Facts offer interesting and little-known bits of information about the United States and Canada and spotlight geographical records and extremes.

Whether you are an armchair traveler or one in earnest, the *Pictorial Atlas of the United States and Canada* offers a clear look at the beauty, variety, history, and spirit of these two countries. Enjoy!

THE UNITED STATES OF AMERICA

*O say, can you see,
by the dawn's early light,*

*What so proudly we hail'd
at the twilight's last gleaming?*

*Whose broad stripes and bright stars,
thro' the perilous fight,*

*O'er the ramparts we watch'd,
were so gallantly streaming?*

*And the rockets' red glare,
the bombs bursting in air,*

*Gave proof thro' the night
that our flag was still there.*

*O say, does that star-spangled
banner yet wave*

*O'er the land of the free
and the home of the brave?*

On September 13, 1814, Francis Scott Key watched British ships attack Fort McHenry, one of the forts defending Baltimore during the War of 1812. The next morning, his joy at seeing the American flag still flying above the fort inspired him to write a poem commemorating the occasion. It was first published under the title, "Defense of Fort McHenry," and later as, "The Star Spangled Banner." The poem gained wide popularity sung to the tune, "To Anacreon in Heaven," the origin of which is uncertain. "The Star Spangled Banner" was officially made the national anthem by Congress in 1931.

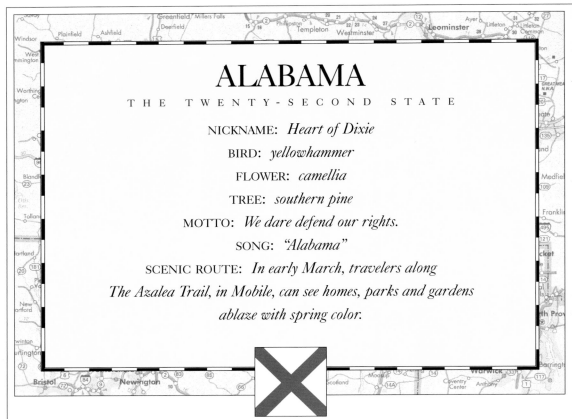

ALABAMA

THE TWENTY-SECOND STATE

NICKNAME: *Heart of Dixie*

BIRD: *yellowhammer*

FLOWER: *camellia*

TREE: *southern pine*

MOTTO: *We dare defend our rights.*

SONG: *"Alabama"*

SCENIC ROUTE: *In early March, travelers along The Azalea Trail, in Mobile, can see homes, parks and gardens ablaze with spring color.*

ALABAMA was named after the Alibamu Indian tribe, which once lived in that region. Because of its central location among the Southern states and its key role in the history of the South, Alabama is nicknamed the "Heart of Dixie." Montgomery, the capital of Alabama, was the first capital of the Confederacy. Confederate President Jefferson Davis took his oath of office on February 18, 1861, at Montgomery's capitol building. A brass star on the steps shows the place where he stood. On March 4, 1861, the Confederate flag was flown, for the first time, over that building. The Stars and Bars, as the flag was called, was designed and made by two women from Marion, Alabama.

Alabama is the only state that has all of the resources needed to make iron and steel. Large deposits of iron ore, coal, and limestone are found there. On Red Mountain, a huge statue of Vulcan, the Roman god of fire and metalworking, watches over nearby Birmingham's iron and steel factories.

For years, cotton was the major crop in Alabama. But in the early 1900s, the boll weevil began to destroy Alabama's cotton plants. Farmers were forced to begin growing other crops. Because of the experiments of scientist George Washington Carver, soybeans and peanuts soon became big cash crops in the South. Carver developed more than 300 products made from peanuts. In the George Washington Carver Museum at Tuskegee University, visitors can tour Carver's laboratory and see his exhibits, as well as those of others.

Huntsville is nicknamed "Rocket City, U.S.A." Many rockets and satellites are designed at the George C. Marshall Space Flight Center there. Visitors to the Alabama Space and Rocket Center can see the world's largest display of missiles and other space equipment. The center also operates a children's space camp.

Vacationers enjoy the sandy beaches along Alabama's Gulf of Mexico shore. The city of Mobile, located on Mobile Bay, just off the Gulf of Mexico, is known for its famous Azalea Trail. In early March, travelers on this 35-mile (56-kilometer) automobile route can see homes, parks, and gardens ablaze with spring color.

For more information about Alabama call (800) ALABAMA (252-2262).

Left, statue of Vulcan on Red Mountain; *far left,* shrimp boats on Bon Secour Bay; *top left,* Cahaba River; *top,* state capitol in Montgomery.

THE name *Alaska* was taken from a Native American word meaning "great land." Alaska is, by far, the largest state in the United States. It is almost a fifth as big as all the rest of the states put together, but it has a smaller population than any other state except Vermont and Wyoming. Alaska is known as the "Last Frontier" because much of the state has not yet been settled completely.

Alaska is a huge peninsula that is linked to Canada. Nearly 500 miles (805 kilometers) of Canadian territory lie between Alaska and the lower 48 states. In 1867, U.S. Secretary of State William H. Seward purchased Alaska from Russia for $7.2 million. This amounted to about two cents per acre of land, or five cents per hectare. At the time, many people claimed that Alaska was a huge wasteland, and they labeled it "Seward's Folly." Since then, Alaska has proved to be a very valuable state with rich resources of lumber, natural gas, gold, coal, and especially oil. In 1977, the Trans-Alaska Pipeline went into service, helping to carry oil from Prudhoe Bay to the port of Valdez. This pipeline is 800 miles (1,300 kilometers) long. Alaska is also the number one state for commercial fishing, with huge amounts of salmon, crab, shrimp, and halibut caught there.

Alaska has stretches of beautiful wilderness and majestic mountains. Mount McKinley, the highest peak in North America, juts 20,320 feet (6,194 meters) into the air. Most of America's active volcanoes are also found in Alaska. Thousands of glaciers, formed by the tremendous snows of the area, are located in Alaska's mountain valleys. One-third of the state is north of the Arctic Circle. At Point Barrow, the northernmost town in the United States, the sun does not set for almost three months in summer and does not rise for two months in winter.

Most of Alaska's population is centered in and around the cities of Anchorage and Fairbanks. The capital city of Juneau is located on Alaska's Panhandle near beautiful Mendenhall Glacier. Farther south, visitors to Ketchikan can see a large collection of totem poles carved by Native Americans. Along the Arctic coast, many Inuit, or Eskimos, still live much as they did hundreds of years ago.

For more information about Alaska call (907) 465-2010.

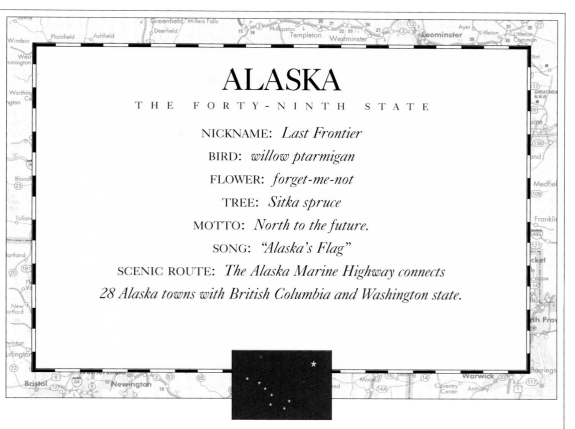

ALASKA
THE FORTY-NINTH STATE

NICKNAME: *Last Frontier*

BIRD: *willow ptarmigan*

FLOWER: *forget-me-not*

TREE: *Sitka spruce*

MOTTO: *North to the future.*

SONG: *"Alaska's Flag"*

SCENIC ROUTE: *The Alaska Marine Highway connects 28 Alaska towns with British Columbia and Washington state.*

Top, Inuit children; *clockwise from lower left,* caribou bull and Mount McKinley; totem carver; Mendenhall Glacier, Juneau; dogsled on the Iditarod Trail.

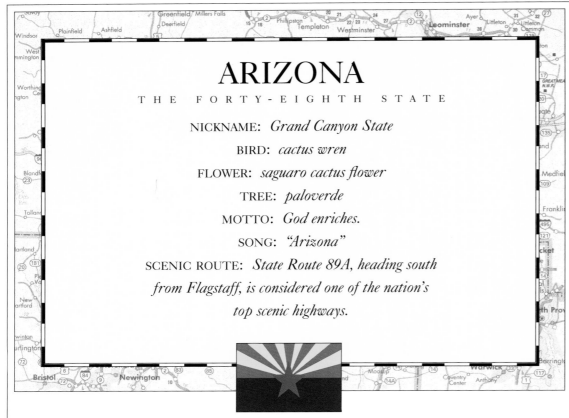

ARIZONA

THE FORTY-EIGHTH STATE

NICKNAME: *Grand Canyon State*

BIRD: *cactus wren*

FLOWER: *saguaro cactus flower*

TREE: *paloverde*

MOTTO: *God enriches.*

SONG: *"Arizona"*

SCENIC ROUTE: *State Route 89A, heading south from Flagstaff, is considered one of the nation's top scenic highways.*

THE warm climate and beautiful scenery of Arizona attract many visitors and new residents each year. Much of the state is desert. At the Arizona-Sonora Desert Museum near Tucson, visitors will find displays of desert animals and plants. The Painted Desert, in north-central Arizona, has miles of beautiful blue, yellow, red, pink, and tan sands and rocks. Navajo Indians used these colorful sands to make sand paintings. Nearby, the Petrified Forest National Park has many ancient logs that have turned to stone.

The name *Arizona* was taken from the Native American word *arizonac*. Its exact meaning is not known. Native American dwellings built more than a thousand years ago can be found in many parts of the state. These ruins are among Arizona's greatest treasures, and some are open to visitors. The cliff dwellings built high in the sandstone walls at Montezuma Castle National Monument, near Camp Verde, are especially well preserved. Today, Arizona has the nation's third-largest Native American population. Nineteen reservations cover almost one-fourth of the state's land. Some of the Native Americans still follow the lifestyles and keep the traditions of their ancestors.

Phoenix, Arizona's capital and largest city, is a center for manufacturing and tourism. More than half of the state's people live in and around Phoenix. Tombstone, located in southeastern Arizona, was a tough mining town in the late 1800s. There, Wyatt Earp fought in a famous gunfight at the O.K. Corral. The town has been restored to look as it did in the days of the Wild West.

Arizona's Grand Canyon is one of the great natural wonders of the world. It measures about 277 miles (446 kilometers) long, up to 18 miles (29 kilometers) wide, and more than a mile (1.6 kilometers) deep. The layers visible on the walls of the Grand Canyon tell the story of the earth's formation. Visitors can hike or ride mules down to the Colorado River at the bottom of the canyon, where the Havasupai Indians have lived for hundreds of years.

For more information about Arizona call (800) 842-8257.

Clockwise from lower left, the north rim of the Grand Canyon; ruins of Native American dwellings in Canyon de Chelly National Monument; Monument Valley; San Xavier; Tonto Natural Bridge State Park; Havasupai Canyon; *top,* desert sunset.

ARKANSAS is split diagonally from northeast to southwest into two triangular regions, the Highlands and the Lowlands. The Ozark and Ouachita mountains stretch across the Highlands in northern and western Arkansas. At the Ozark Folk Center in Mountain View, quilters, weavers, woodcarvers, and others demonstrate their crafts. Local musicians also entertain visitors with their fiddle music played on handmade instruments. Nearby Blanchard Springs Caverns has beautiful stalactites and stalagmites.

The Mississippi River forms Arkansas' eastern border. The name Arkansas was taken from a Native American word meaning "downstream people." Rice fields are plentiful in the wet areas of this low region. Arkansas is the number one state for rice production and is also a leader in soybean and cotton production. Rice farmers sometimes flood unused fields and raise fish in them. Because of its many rich natural resources and growing industries, Arkansas is nicknamed the "Land of Opportunity."

Arkansas' capital and largest city is Little Rock. It was founded in 1820 and named Little Rock because of its location on the smaller of two rock banks on the Arkansas River. Little Rock and Fort Smith are Arkansas' main centers for manufacturing and trade.

Arkansas is famous for its bubbling springs of water. Mammoth Spring in the Ozarks is one of the largest springs in the world. Each day about 865 million gallons (3.3 billion liters) of cold water pour out of the earth. Other springs in Arkansas have hot mineral waters. Many people believe these waters have healing powers. Long ago, Native Americans considered the location at Hot Springs, in the Ouachita Mountains, to be holy ground. Warring tribes laid down their weapons and bathed together there. Today, these hot waters are piped into bathhouses for the enjoyment of visitors.

The Crater of Diamonds State Park near Murfreesboro is the only public diamond field in North America. For a fee, visitors can dig for diamonds in the 78-acre (32-hectare) field and keep whatever they find. Thousands of people have found diamonds of good size and value.

For more information about Arkansas call (800) NATURAL (628-8725).

ARKANSAS

THE TWENTY-FIFTH STATE

NICKNAME: *The Land of Opportunity*

BIRD: *mockingbird*

FLOWER: *apple blossom*

TREE: *pine*

MOTTO: *The people rule.*

SONG: *"Arkansas"*

SCENIC ROUTE: *Arkansas' first scenic byway, Ark. 7, has been selected by numerous experts as one of the Top 10 scenic drives in America.*

Top, Whitaker's Point, Ozark National Forest; *right,* Buffalo National River; *far right,* the state capitol in Little Rock; *top right,* The Old Mill in North Little Rock.

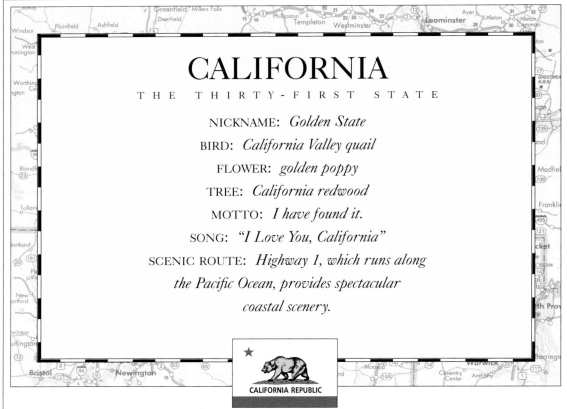

CALIFORNIA

THE THIRTY-FIRST STATE

NICKNAME: *Golden State*

BIRD: *California Valley quail*

FLOWER: *golden poppy*

TREE: *California redwood*

MOTTO: *I have found it.*

SONG: *"I Love You, California"*

SCENIC ROUTE: *Highway 1, which runs along the Pacific Ocean, provides spectacular coastal scenery.*

CALIFORNIA REPUBLIC

EARLY explorers from Spain named California after a treasure island in a popular sixteenth-century Spanish novel. California has been the fastest-growing state since it entered the Union in 1850. More people live there than in any other state. It also has the country's second largest population of Native Americans.

California is sometimes called the "Golden State." This nickname refers not only to California's 1849 gold rush, but also to the state's sunny weather. California's Central Valley is a prime farming area, and the state ranks first in total food production. One-third of America's canned and frozen fruits and vegetables are grown there. Sacramento, the state capital, is located on the Sacramento River in the heart of this farming region.

California is a large state of great variety and beauty. Much of southeastern California is desert. Death Valley, lying 282 feet (86 meters) below sea level, is the lowest point in the Western Hemisphere. The Sierra Nevada, in eastern California, is the highest mountain range in the continental United States. Beautiful waterfalls and steep mountains in Yosemite National Park offer spectacular views. Farther north along the coastline are forests of beautiful redwoods. These trees grow to heights of 300 feet (91 meters) and are the tallest living things on earth.

California has 840 miles (1,352 kilometers) of coastline along the Pacific Ocean. The state is a leader in commercial fishing, with huge catches of tuna and crab. All of the state's largest cities are along the coast. San Diego, near the border of Mexico, has one of the best zoos in the world. Los Angeles is California's biggest city, and visitors there can tour Universal Studios to see how movies are made. Nearby Marineland is the world's largest oceanarium and houses a variety of porpoises, sharks, and other sea animals. Disneyland, in Anaheim, is one of the most popular amusement parks in the United States.

Farther north is San Francisco, the "City of Hills." Cable cars carry passengers up and down the steep streets above the bay. Many people enjoy driving on Lombard Street—the most crooked street in the world!

For more information about California call (800) TO CALIF (862-2543).

Clockwise from lower left, Death Valley; San Francisco; Disneyland; Venice Beach; Santa Barbara mission; *top*, Heavenly Valley, Lake Tahoe; *opposite page*, Gull Rock, Sonoma.

14

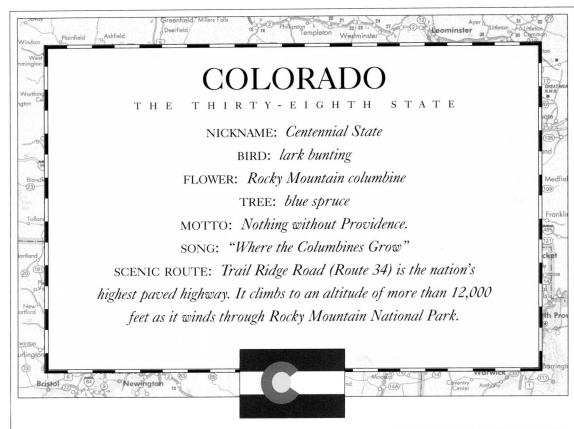

COLORADO

THE THIRTY-EIGHTH STATE

NICKNAME: *Centennial State*

BIRD: *lark bunting*

FLOWER: *Rocky Mountain columbine*

TREE: *blue spruce*

MOTTO: *Nothing without Providence.*

SONG: *"Where the Columbines Grow"*

SCENIC ROUTE: *Trail Ridge Road (Route 34) is the nation's highest paved highway. It climbs to an altitude of more than 12,000 feet as it winds through Rocky Mountain National Park.*

THE poem "America the Beautiful," which eventually became a song, was inspired by a breathtaking view from Pike's Peak in Colorado. There are plenty of such views in this state of great natural beauty. Wheat fields and cattle ranches in the east give way to the majestic Rocky Mountains in the west. More than 50 peaks rise higher than 14,000 feet (4,270 meters) above sea level.

Colorado is the highest state above sea level. Its capital and largest city, Denver, is known as "Mile High City" because it is exactly one mile (1,609 meters) above sea level. There, the United States Mint produces millions of coins each year.

Colorado means "colored red" in Spanish. The state was named after the Colorado River, which flows through red sandstone canyons. Colorado is known as the "Centennial State" because it joined the Union in 1876—100 years after the United States was born.

Each year people travel to Colorado to sightsee, hike, camp, or ski. In the summer, an old-time locomotive carries passengers on a 90-mile (140-kilometer) round trip through the canyons between Durango and Silverton. Visitors to Royal Gorge, near Canon City, can walk on the highest suspension bridge in the world. It stands 1,053 feet (321 meters) over a canyon carved by the Arkansas River.

At the U.S. Air Force Academy, located in the foothills of the Rocky Mountains just north of Colorado Springs, cadets attend college and receive officer training. A nearby scenic park, Garden of the Gods, is famous for its huge red sandstone masses that form strange shapes and patterns.

In the eleventh and twelfth centuries, Native Americans in southwestern Colorado made great adobe and rock houses that resembled today's apartment buildings. For protection against enemies, many of the houses were built along steep canyon walls and on the edges of cliffs. Today, a number of these high cliff dwellings still stand in Mesa Verde National Park. The largest cliff house, Cliff Palace, has over 200 rooms, in which as many as 400 people lived at one time.

For more information about Colorado call (800) COLORAD (265-6723).

Clockwise from lower left, Denver; Spruce Tree House, Mesa Verde National Park; Maroon Bells near Aspen; Civic Center Park in Denver; Sylvan Lake State Park; State Forest Recreational Area, Mencos; *top,* rafting in Lory State Park.

THE name *Connecticut* was taken from the Algonquian Indian word *quinnehtukqut*, meaning "on the long tidal river." The 410-mile (660-kilometer) Connecticut River splits the state in half and flows into Long Island Sound. Sandy beaches along the Sound are popular vacation spots. Fine natural harbors have also made Connecticut a center for shipbuilding and sailing. Mystic Seaport, a thriving waterfront village in the country's early days, has been restored to look as it would have appeared in the 1840s. Sailors dressed in period garb work on wooden whalers and clipper ships anchored there.

Nearby is Groton, the submarine capital of the world. The Submarine Library and Museum there traces the history of submarines and displays models of famous ones. The first nuclear-powered submarine, the U.S.S. *Nautilus*, was built at Groton in 1954. After years of service, the *Nautilus* is now docked there and open for tours. The U.S. Coast Guard Academy is across the Thames River in New London.

At the Constitutional Convention of 1787, Connecticut delegates offered a plan called the "Connecticut Compromise," which helped solve the problem of how many people would be elected to Congress from each state. For this reason, Connecticut became known as the "Constitution State."

In 1798, Eli Whitney changed industry forever when he built a factory where guns were made by machines quickly and uniformly. This was the beginning of mass production; previously, all products in the United States were made by hand. Factories that make clocks, bicycles, and machine tools were started in Connecticut and are still thriving businesses today. Much of this industry is centered around Bridgeport, Connecticut's largest city.

The insurance industry began in the riverside village of Hartford, when ship owners banded together to lessen the risks of long voyages at sea. Today, Hartford is the state capital and second-largest city. It is known as "Insurance City" because so many insurance companies have their headquarters there.

For more information about Connecticut call (800) CT BOUND (282-6863).

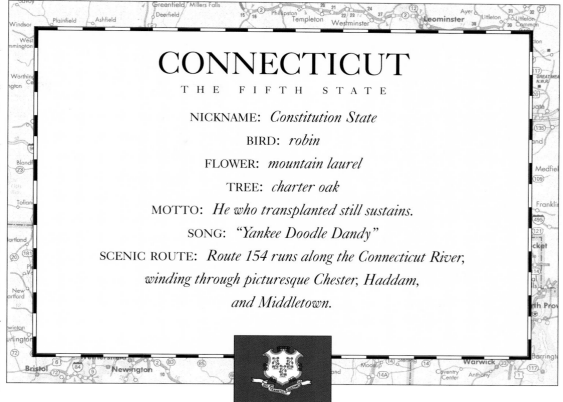

CONNECTICUT
THE FIFTH STATE

NICKNAME: *Constitution State*

BIRD: *robin*

FLOWER: *mountain laurel*

TREE: *charter oak*

MOTTO: *He who transplanted still sustains.*

SONG: *"Yankee Doodle Dandy"*

SCENIC ROUTE: *Route 154 runs along the Connecticut River, winding through picturesque Chester, Haddam, and Middletown.*

Top, Suffield highway; *right,* Mark Twain's house, Hartford; *far right,* center commons, Stonington; *top right,* Mystic Seaport.

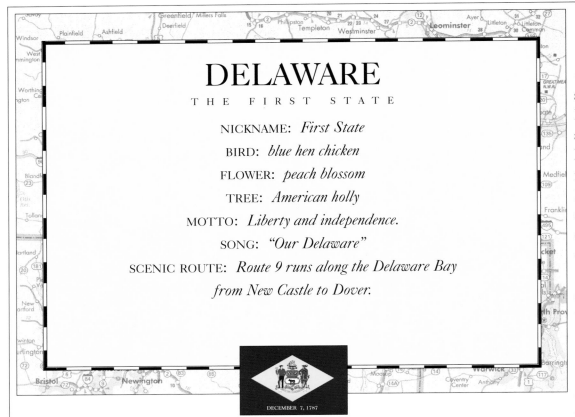

DELAWARE

THE FIRST STATE

NICKNAME: *First State*

BIRD: *blue hen chicken*

FLOWER: *peach blossom*

TREE: *American holly*

MOTTO: *Liberty and independence.*

SONG: *"Our Delaware"*

SCENIC ROUTE: *Route 9 runs along the Delaware Bay from New Castle to Dover.*

DECEMBER 7, 1787

DELAWARE is only 35 by 96 miles (56 by 154 kilometers) in size, the second-smallest state after Rhode Island. Much of Delaware is located on the Delmarva Peninsula, a large body of land shared by Delaware, Maryland, and Virginia. On the flat, fertile farmland there, many farmers raise broilers. Poultry is the number one farm product, but soybeans, corn, and mushrooms are also leading crops. Dover, the state capital, is a center for commerce in this farming area.

Delaware is known for its long stretch of sandy beaches located along the Atlantic shore from Lewes to Fenwick Island. Rehoboth, located between the two, is sometimes called the "Nation's Summer Capital" because so many members of Congress and government officials vacation there to escape the hot weather of Washington, D.C. Visitors to the Zwaanendael Museum in Lewes can learn about early Indians and the history of Lewes' seaport. The Cape May-Lewes Ferry crosses the Delaware Bay several times each day and carries passengers and cars between Lewes, Delaware, and Cape May, New Jersey.

Much of Delaware's history has been shaped by the du Pont family. Around 1800, Frenchman Éleuthère Irénée du Pont and his family moved to America and settled near Wilmington. They built and operated a gunpowder mill on Brandywine Creek. The business was very successful, and generations of the du Pont family went on to establish the world's largest chemical research company. Today, Wilmington is known as the "Chemical Capital of the World." Much of the original du Pont gunpowder mill and other exhibits of early American industry are now on display at the Hagley Museum. Nearby Winterthur Museum houses a du Pont collection of Early American furniture in a beautiful mansion of more than 100 rooms.

Delaware was named after the British governor of Virginia, Lord De La Warr. It is known as the "First State" because it approved the U.S. Constitution on December 7, 1787, the first state to do so.

For more information about Delaware call (800) 441–8846.

Left, the Cecil Bedroom at the Winterthur Museum; *far left;* Brandywine Valley; *top left,* Hagley Museum; *top,* Rehoboth Beach.

18

THE "Sunshine State" has warm weather almost all year round. Its mild, wet climate provides ideal conditions for growing more than 3,000 different varieties of flowers. The word *Florida* means "full of flowers" in Spanish. Sunny Florida is also the number one producer of oranges and grapefruit in the United States.

Florida is a large peninsula, about 450 miles (725 kilometers) in length, with huge stretches of coastline. Its many beaches are popular vacation areas. Miami Beach alone attracts more than two million visitors each year. Seashell lovers go to Sanibel Island in the Gulf of Mexico to collect different specimens on its beaches. The capital city of Tallahassee is located in northwestern Florida, near the Georgia border. Jacksonville, in the northeastern part of the state, is Florida's biggest city. Its city limits include a larger area than any other U.S. city except Juneau, Alaska.

At the southern tip of Florida, hundreds of islands known as the Florida Keys help shield the beaches from the Atlantic Ocean's might. They are a scuba diver's dream. Near Key Largo is John Pennekamp Coral Reef State Park, the first undersea park on the mainland of the United States. Some 30 different types of coral can be seen in the park's crystal-clear water.

The unspoiled swamplands of southern Florida's Everglades National Park are home to an amazing variety of wildlife. Alligators and beautiful birds are among the many animals living there. Seminole Indians fled to the Everglades during the Second Seminole War (1835–42), and many live there today in a village on the Tamiami Trail.

St. Augustine, in northern Florida, is the oldest city in the United States. It was founded by Spanish settlers in 1565, and visitors today can walk through narrow streets built hundreds of years ago.

In 1981, the first manned space shuttle launches were made from Kennedy Space Center, located in Cape Canaveral. The Space Center, which conducts bus tours for visitors, continues to play a large part in NASA development.

Walt Disney World, the most popular amusement park in the world, is located near Orlando in central Florida. Visitors there can enjoy the Magic Kingdom, EPCOT Center, and the Disney-MGM Studios Theme Park.

For more information about Florida call (904) 487-1462.

FLORIDA

THE TWENTY-SEVENTH STATE

NICKNAME: *Sunshine State*

BIRD: *mockingbird*

FLOWER: *orange blossom*

TREE: *sabal palmetto palm*

MOTTO: *In God we trust.*

SONG: *"Old Folks at Home"*

SCENIC ROUTE: *Route 699 provides a beautiful look at Florida's diverse west coast.*

Top, Fort Myers; *clockwise from lower left,* Miami; Walt Disney World; Saint Augustine; Florida's vibrant flamingos.

GEORGIA leads all other states in growing "goobers" (a regional name for peanuts). Georgia's warm climate, rich soil, and long, wet summers work together to produce more than 1.5 billion pounds (680,388,555 kilograms) of peanuts each year. About half of that crop is made into peanut butter. Jimmy Carter, a former Georgia governor and the thirty-ninth president of the United States, raised peanuts on a farm in Plains. Georgia is also a leading producer of pecans, peaches, tobacco, and poultry.

Seventy percent of Georgia is forested. The northern part of the state has mainly pine and hardwood trees, while pine and live oak trees are found in the south. The highest mountain peak in the state is Brasstown Bald Mountain, rising 4,784 feet (1,458 meters) above sea level near the northern border. Springer Mountain, the southernmost peak of the Blue Ridge Mountains, marks the end of the Appalachian Trail. Along the Atlantic coast are hundreds of small islands; several of them are popular vacation resorts.

Atlanta, Georgia's capital and largest city, is home to the Martin Luther King Jr. Center. Visitors can view King's grave and learn about ongoing efforts to promote peace and racial harmony. Near Atlanta is Stone Mountain, the world's largest granite boulder. There, the figures of Civil War Confederate leaders Jefferson Davis, Robert E. Lee, and Stonewall Jackson are carved into the mountainside.

Columbus, Georgia's second-largest city, is called the "Fountain City" because of its many beautiful fountains. The city of Savannah is an important port near the Atlantic coast.

The Okefenokee Swamp covers a large section of southeastern Georgia and northern Florida. This big, freshwater swamp is a refuge for alligators and other wildlife. Visitors can travel the water trails on guided boat tours to view beautiful cypress trees and hanging Spanish moss.

Named after England's King George II, Georgia was the last of the original 13 colonies to be founded. In 1788, it became the fourth state of the Union. No state east of the Mississippi River is larger than Georgia. Because of its size and importance, Georgia is known as the "Empire State of the South."

For more information about Georgia call (800) VISIT GA (847-4842).

Top, Underground Atlanta; *clockwise from lower left,* Forsyth Park, Savannah; George Woodruff House, Macon; Martin Luther King Jr. memorial; alligator in Okefenokee swamp; *opposite page,* Stone Mountain Monument.

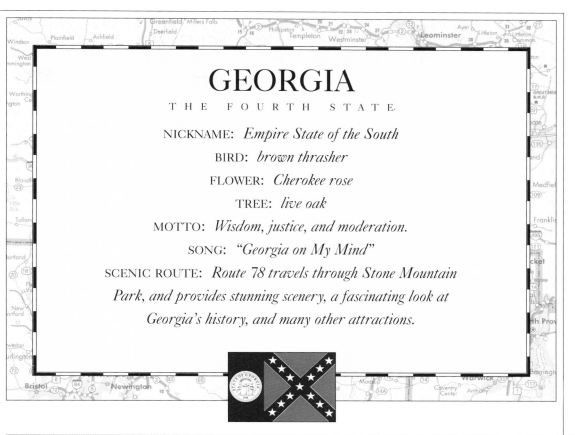

GEORGIA
THE FOURTH STATE

NICKNAME: *Empire State of the South*
BIRD: *brown thrasher*
FLOWER: *Cherokee rose*
TREE: *live oak*
MOTTO: *Wisdom, justice, and moderation.*
SONG: *"Georgia on My Mind"*
SCENIC ROUTE: *Route 78 travels through Stone Mountain Park, and provides stunning scenery, a fascinating look at Georgia's history, and many other attractions.*

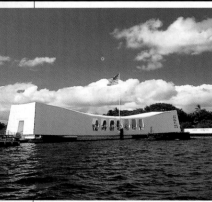

HAWAII

THE FIFTIETH STATE

NICKNAME: *Aloha State*

BIRD: *Hawaiian goose*

FLOWER: *yellow hibiscus*

TREE: *kukui*

MOTTO: *The life of the land is perpetuated in righteousness.*

SONG: *"Hawaii Ponoi" ("Hawaii's Own")*

SCENIC ROUTE: *Chain of Craters Road, in Hawaii Volcanoes National Park on the island of Hawaii, passes enormous volcanic craters and a black sand beach as it winds toward the coast.*

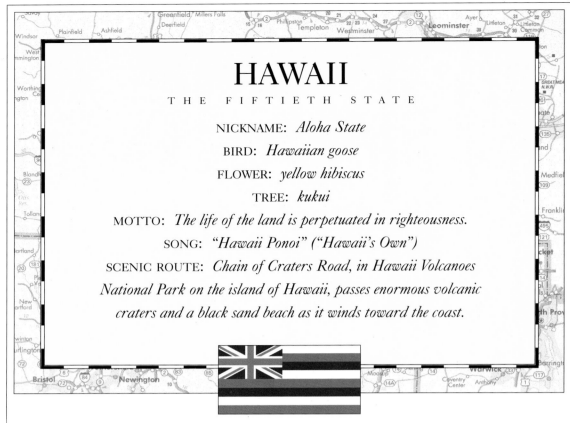

VISITORS to Hawaii are often greeted with beautiful leis, strings of flowers that can be worn around the neck. Because of the warm hospitality of the Hawaiian people, Hawaii is called the "Aloha State." In Hawaiian, *aloha* means "love."

Hawaii is the only state not on the continent of North America. It is a group of 132 islands stretching 1,523 miles (2,451 kilometers) in the middle of the Pacific Ocean. The islands were all created by repeated volcanic eruptions. One hundred twenty-four of the islands are tiny and without a permanent population. Most of Hawaii's people live on seven of the eight main islands at the southeastern end of the chain. With a year-round mild climate and beautiful plant life, the islands are considered by many to be a tropical paradise.

The largest of the islands is Hawaii, for which the entire chain is named. This island has two active volcanoes, Mauna Loa and Kilauea. From roped-off ledges around the craters, bubbling lava and fire fountains can be seen. People enjoy swimming and scuba diving off the sparkling black sands of Kaimu Black Sand Beach.

The islands of Lanai, Maui, and Molokai have many sugarcane and pineapple plantations. Hawaii is the number one producer of pineapples in the United States. Beautiful Waimea Canyon, on Kauai, is known as the "Grand Canyon of Hawaii." This colorful gorge is 2,000 feet (610 meters) deep. The island of Niihau is privately owned and cannot be visited without special permission. Kahoolawe is used by the U.S. Armed Forces for target practice, but no people live there.

Pearl Harbor Naval Base is on the island of Oahu. Japanese warplanes attacked Pearl Harbor on December 7, 1941, drawing the United States into World War II. A memorial is built over the U.S.S. *Arizona*, which was sunk during the attack. Honolulu, Hawaii's capital and largest city, and Waikiki Beach are also located on Oahu.

The original settlers of Hawaii were Polynesians. They traveled in big canoes from other islands in the Pacific Ocean. The word *Hawaii* might have come from a Polynesian chief's name, "Hawaii-loa," or from "Hawaiki," the name of a Polynesian land that is the subject of legend.

For more information about Hawaii call (808) 923–1811.

Clockwise from lower left, Kaluakoi, Molokai; U.S.S. *Arizona* War Memorial; Kilauea volcano, Kalapana; surfing; Iao Valley, Maui; *top,* hula dance, Oahu.

IDAHO is a scenic area with 22 mountain ranges, many of them very rugged. The Northern Rocky Mountains cover most of the northern part of the state, and the Middle Rockies are in the southeastern part of the state. Some people from Idaho boast that if their state were ironed out flat, it would be as big as Texas. Camping, hiking, fishing, and hunting are popular activities throughout the area. People also enjoy skiing and ice skating in such beautiful areas as Sun Valley in southern Idaho.

When Idaho became the forty-third state in 1890, it took its name from the Idaho Territory, which included the land of Idaho state and other surrounding areas. Idaho's capital and largest city is Boise. It is located on the Boise River in the southwestern part of the state.

Along the border with Oregon is Hells Canyon, the deepest chasm in North America. The powerful Snake River has carved out depths of up to 7,900 feet (2,408 meters) there. This chasm is deeper than the Grand Canyon in Arizona.

Southern Idaho has rich farmland, where huge crops of potatoes have made Idaho the number one producer of potatoes in the United States. Wheat, sugar beets, hay, and barley also rank high in production.

Idaho is an important mining state. It is nicknamed the "Gem State" because of its many valuable minerals. Throughout the state, ghost towns stand as reminders of the gold and silver rushes of the 1860s. One of the best-known ghost towns is Silver City in Owyhee County. Today, Idaho has large deposits of phosphate, lead, and gold, and it is the country's leading producer of silver.

There are hundreds of underground caverns in Idaho. One of the more spectacular is Crystal Ice Cave near Pocatello. There, you can see interesting ice and stone formations as well as a frozen waterfall and frozen river.

At Craters of the Moon National Monument in southern Idaho, visitors can walk on what appears to be a barren lunar landscape. Many years ago, tremendous eruptions occurred and left miles of colorful lava hardened into cones, bridges, and other strange shapes. This area is one of the best examples of volcanic action ever seen on earth.

For more information about Idaho call (800) 635-7820

IDAHO
THE FORTY-THIRD STATE

NICKNAME: *Gem State*
BIRD: *mountain bluebird*
FLOWER: *syringa*
TREE: *western white pine*
MOTTO: *It is perpetual.*
SONG: *"Here We Have Idaho"*
SCENIC ROUTE: *Route 95 is a varied and picturesque route that travels along the borders of Washington and Oregon, passing through several Indian reservations.*

Top, skiing in Sun Valley; *clockwise from lower left,* Teton mountains; hikers on the Centennial Trail; Shoshone Falls; hotel in Silver City.

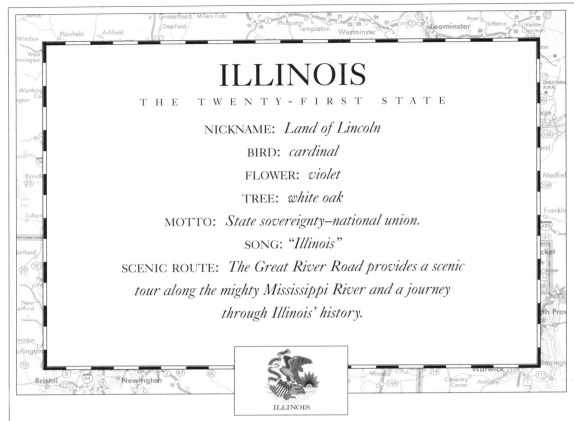

ILLINOIS

T H E   T W E N T Y - F I R S T   S T A T E

NICKNAME: *Land of Lincoln*

BIRD: *cardinal*

FLOWER: *violet*

TREE: *white oak*

MOTTO: *State sovereignty–national union.*

SONG: *"Illinois"*

SCENIC ROUTE: *The Great River Road provides a scenic tour along the mighty Mississippi River and a journey through Illinois' history.*

ILLINOIS

ILLINOIS is the "Land of Lincoln." Abraham Lincoln moved there at the age of 21 and always considered it to be his home. At New Salem State Park, the town of New Salem has been completely restored to look as it did when Lincoln lived there from 1831 to 1837. Lincoln's later home in Springfield, the state capital, is a national historic site. The sixteenth president of the United States is buried with his wife and three of their four sons at Oak Ridge Cemetery in Springfield.

Illinois was named after the Illini Indian tribe. This tribe was one of many that lived on the land before white settlers arrived. Thousands of years earlier, prehistoric Native Americans lived and hunted by the Great Lakes. They were known as Mound Builders because of the large mounds of earth they shaped for worship and burial. Several thousand of these mounds can be seen throughout Illinois. At the Dickson Mounds Park near Lewistown, visitors can tour a museum of Stone Age relics and see the contents of a mound that was partially dug open.

Most of Illinois is flat, scraped by glaciers years ago. The Mississippi Palisades in the northwest and the Illinois Ozarks in the south were missed by the glaciers, however. There, the land is steep and hilly. Illinois has rich, fertile farmland, and is the number one producer of soybeans in the United States. Other major crops include corn, oats, and wheat.

Chicago is Illinois' largest city and the greatest inland port in the world. The Illinois Waterway links the St. Lawrence Seaway and Great Lakes to the Mississippi River and Gulf of Mexico. This provides an important shipping route to the heart of America. The Chicago-O'Hare International Airport is one of the busiest airports in the world. Approximately 1,500 aircraft fly in or out of there each day. Chicago is also where skyscrapers were invented. The tallest office building in the world, Chicago's Sears Tower, is 1,454 feet (443 meters) high. From the observation deck on the 103rd floor, you can see Wisconsin.

For more information about Illinois call (800) 223-0121

Left, Lincoln's childhood home; *upper left,* Chicago; *top,* state capitol in Springfield.

INDIANA was once home to many Indian tribes, and great Indian leaders such as Miami Chief Little Turtle and Shawnee Chief Tecumseh played key roles in the state's history. The name *Indiana* means "land of the Indians." Today, only museum exhibits and Native American names such as Kokomo, Wawasee, Kankakee, and Miami are left as reminders of what came before.

Indiana is known as the "Hoosier State." It is said that Indiana pioneers often called out, "Who's here?" when new settlers arrived. With wave after wave of newcomers, the words might have been slurred, resulting in "Hoosier."

Abraham Lincoln moved to Indiana in 1816 at the age of seven and lived there until age 21. Visitors can see his original log home at the Lincoln Boyhood National Memorial near Lincoln City. A restored farm from the 1800s is also on the site.

Fort Wayne was founded by General "Mad" Anthony Wayne in 1794 to protect settlers from Indian raids. Wayne was called "Mad" because of his very daring attacks on the British during the Revolutionary War. Visitors to Fort Wayne can see a remade 1816 army fort and watch military demonstrations. John Chapman, better known as Johnny Appleseed, is buried nearby in Johnny Appleseed Park.

Indiana is an important state for farming and industry. Corn, popcorn, and soybeans are Indiana's main crops. The city of Gary, on the shore of Lake Michigan, is known for its many steel mills and oil refineries. Nearby, the Indiana Dunes National Lakeshore offers beautiful scenery and recreational opportunities. As an outgrowth of the metals industry, Indiana is also known for manufacturing cars, trucks, boats, and recreational vehicles.

Race car fans enjoy the Indianapolis Motor Speedway. Many races are run on the 2.5-mile (4-kilometer) oval track located in the state's capital and largest city, Indianapolis. Each Memorial Day weekend, the top drivers in the country compete in the Indianapolis 500, a 500-mile (805-kilometer) race of speed and endurance.

For more information about Indiana call (800) 289-6646

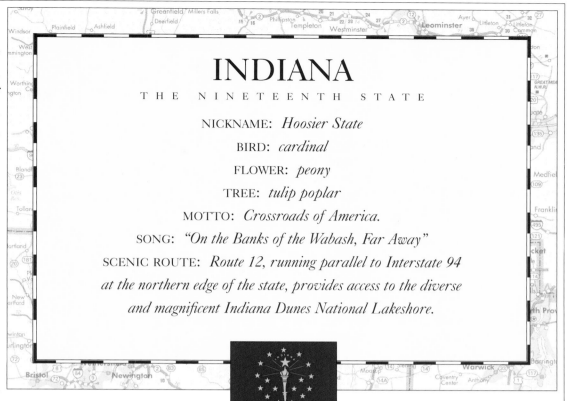

INDIANA
T H E   N I N E T E E N T H   S T A T E

NICKNAME: *Hoosier State*

BIRD: *cardinal*

FLOWER: *peony*

TREE: *tulip poplar*

MOTTO: *Crossroads of America.*

SONG: *"On the Banks of the Wabash, Far Away"*

SCENIC ROUTE: *Route 12, running parallel to Interstate 94 at the northern edge of the state, provides access to the diverse and magnificent Indiana Dunes National Lakeshore.*

Top, Indiana Dunes; *clockwise from lower left,* the state capitol in Indianapolis; farm in Dearborn County; the Indianapolis 500; the Ohio River.

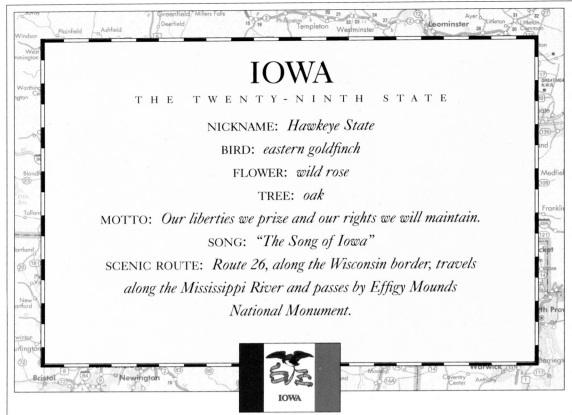

IOWA

THE TWENTY-NINTH STATE

NICKNAME: *Hawkeye State*

BIRD: *eastern goldfinch*

FLOWER: *wild rose*

TREE: *oak*

MOTTO: *Our liberties we prize and our rights we will maintain.*

SONG: *"The Song of Iowa"*

SCENIC ROUTE: *Route 26, along the Wisconsin border, travels along the Mississippi River and passes by Effigy Mounds National Monument.*

IOWA's plentiful rainfall and fertile soil make it one of America's top farming states. Almost 95 percent of Iowa's land is used for farming. Iowa often leads all other states in the growing of corn, and stalks as high as 30 feet (9 meters) have been measured. Iowa also raises more hogs than any other state.

Farming has long been a way of life in Iowa. At the Living History Farms near Des Moines, the state's capital and largest city, visitors can learn about farming in Iowa before 1900. A farm of the future is also on display. Each August the Iowa State Fair, one of the largest fairs in the United States, is held in Des Moines.

Iowa was named after the Iowa Indians who once lived in the area. It was nicknamed the "Hawkeye State" after the famous Indian Chief Black Hawk. Black Hawk led an uprising against settlers living west of the Mississippi River. The U.S. Army defeated Black Hawk and gained possession of a tract of land that later became part of Iowa. Today, the Mesquakie Indians live on a reservation in Tama. Visitors can see the beautiful baskets, beadwork, and leather products they make.

Southwest of Cedar Rapids are the Amana Colonies. This settlement of seven villages was begun in 1855 by a group of people who banded together to live a simple life of shared work and rewards. Although it is no longer a cooperative effort, the businesses are still successful today. Villagers live simply, make crafts, and operate restaurants serving homemade foods.

Herbert Hoover, the thirty-first president of the United States, lived in West Branch in eastern Iowa. His home and grave site are surrounded by a beautiful park, and a library and museum are open to visitors.

Like Illinois, Iowa was once home to Mound Builders, Native Americans who constructed large piles of earth and stone to use for burial and worship. At the Effigy Mounds National Monument near McGregor, many of these mounds can be seen. Some are shaped like animals and measure 300 feet (91 meters) long.

For more information about Iowa call (800) 345-IOWA (345-4692)

Left, a farm near Des Moines; *upper left,* the state capitol in Des Moines; *top,* Loess Hills in western Iowa.

THE exact geographical center of the continental United States is in Kansas, near the town of Lebanon. From the air Kansas looks like a huge patchwork quilt, with its flat prairie land divided into squares of wheat, corn, and soybeans. As the number one wheat-growing state, Kansas is often called the "Breadbasket of America." It is also known as the "Sunflower State" because of the tall, yellow flowers that thrive on the hot midwestern plains. A section of land running north-south through the Flint Hills is the last vestige of the tall-grass prairie that once covered much of the Midwest.

Kansas is named after the Kansa Indians, who once lived on the land. Several Native American reservations are located in Kansas.

The state's largest city is Wichita, the "Airplane Capital of the World." More civilian airplanes are made there than anywhere else. Farther north is the city of Abilene, once a wild cowtown. The legendary James "Wild Bill" Hickok was the town's marshall in 1870. Abilene was also the home of the thirty-fourth president of the United States, Dwight D. Eisenhower. East of both Wichita and Abilene is Topeka, the state's capital. The Menninger Foundation, with its neuropsychiatric clinic and research labs, is located there.

Southwest of the town of Independence is a remake of the log cabin known as the "Little House on the Prairie." It is built on the site where Laura Ingalls Wilder, who wrote the book, *Little House on the Prairie*, lived as a child. A one-room schoolhouse and rural post office are also there.

Dodge City was once the "Cowboy Capital of the World." Huge herds of cattle were brought there from Texas to be shipped by railway to markets in other states. Today, the city's Front Street has been rebuilt to look as it did in the days of the Wild West, when Wyatt Earp and Bat Masterson lived there. Exciting re-enactments are still staged in the streets of Dodge City.

Raising beef cattle has remained a big business in Kansas, but today's cowboys may be seen herding cattle while riding motorcycles.

For more information about Kansas call (800) C KANSAS (252-6727)

KANSAS

THE THIRTY-FOURTH STATE

NICKNAME: *Sunflower State*

BIRD: *western meadowlark*

FLOWER: *sunflower*

TREE: *cottonwood*

MOTTO: *To the stars through difficulties.*

SONG: *"Home on the Range"*

SCENIC ROUTE: *Route 177 travels through El Dorado Wildlife Area and El Dorado State Park.*

KANSAS

Top, Kansas' state flower, the sunflower; *clockwise from lower left,* the state capitol in Topeka; aerial view of wheat fields; Front Street in Dodge City.

KENTUCKY

THE FIFTEENTH STATE

NICKNAME: *Bluegrass State*

BIRD: *cardinal*

FLOWER: *goldenrod*

TREE: *Kentucky coffee tree*

MOTTO: *United we stand, divided we fall.*

SONG: *"My Old Kentucky Home"*

SCENIC ROUTE: *Route 31E is a beautiful drive that passes by the Lincoln Birthplace National Historic Site, crosses the Blue Grass Parkway–another scenic drive–and continues north to Louisville.*

KENTUCKY stretches from the Appalachian Mountains to the Mississippi River. In between are rolling countrysides covered with grasses that bear blue blossoms in spring. This part of the "Bluegrass State" is horse country. Champion racehorses have been raised and trained there for more than 150 years. Horse lovers can visit the Kentucky Horse Park in Lexington to see a horse farm in operation and view many different breeds of horses. The famous racehorse, Man O' War, is buried there. Each May, the country's fastest horses race in the Kentucky Derby at Churchill Downs in Louisville, the state's largest city.

The name *Kentucky* was taken from a Cherokee Indian word that possibly meant "meadowland." One of Kentucky's earliest settlers was Daniel Boone. He blazed a trail in the wilderness through the Cumberland Gap in 1775, creating Kentucky's first road. Later, Boone founded the city of Boonesborough on the Kentucky River. Fort Boonesborough is open to visitors today. Daniel Boone and his wife, Rebecca, are buried in a cemetery in Frankfort, Kentucky's capital.

During the Civil War, Kentucky was a border state. Native sons fought on the Union and Confederate sides. Oddly enough, President Abraham Lincoln and Confederate President Jefferson Davis were born only one year and 100 miles (161 kilometers) apart from each other in Kentucky. Lincoln's birthplace is a national historic site near Hodgenville in the middle of the state.

Kentucky's Mammoth Cave is the longest cave system in the world. Its more than 200 miles (322 kilometers) of underground passages, as well as its waterfalls and underground rivers and lakes, have not yet been fully explored. Kentucky also has tremendous mineral deposits. Its huge beds of coal make it the number one coal-producing state.

Huge U.S. gold reserves, worth more than $6 billion, are stored in underground vaults at Fort Knox, south of Louisville. During World War II, priceless documents such as the Declaration of Independence, the U.S. Constitution, and Lincoln's Gettysburg Address were also kept in the vaults for safekeeping.

For more information about Kentucky call (800) 225-8747

Clockwise from lower left, Fort Boonesborough State Park; a Kentucky farm scene; Cumberland Falls State Park; *top*, bluegrass horse farm.

LOUISIANA has quiet streams and busy ports, cotton fields and oil wells, beautiful mansions and Dixieland jazz. The state was named after King Louis XIV of France. It is known as the "Pelican State" because of all the brown pelicans that live there. Louisiana is a bird-lover's paradise, and is where John J. Audubon drew many of the pictures for his book, *Birds of America*. Each spring, millions of migrating birds stop on Grand Isle to rest after crossing the Gulf of Mexico.

The mighty Mississippi River loops along and through 596 miles (916 kilometers) of Louisiana before it empties into the Gulf of Mexico. To help prevent flooding, levees line the riverbanks. Steamboats still cruise the waters, often side-by-side with ocean-going ships. At the river's mouth, soil deposits from upriver continuously build a delta. Silt from the Rocky Mountains, black soil from the Dakota Badlands, and dirt from a riverbank in Ohio all blend together to make this fertile farmland.

Early in the nineteenth century, pirate Jean Laffite terrorized people along the Gulf of Mexico coast. It is said that he buried his treasures in the swamps and bayous there. Today, offshore oil rigs sprout up from Gulf waters as signs of a modern-day treasure. Louisiana is second only to Texas in the production of oil and natural gas.

South-central Louisiana is "Cajun Country," an area settled by French Canadians from eastern Canada, especially Nova Scotia. Today, many Cajuns still speak French and fish or trap furs for a living.

New Orleans, Louisiana's largest city, is one of the busiest ports in America. It was settled years ago by French and Spanish colonists. Their descendants, known as Creoles, are famous for their spicy cooking. New Orleans is also home to Mardi Gras, French for "Fat Tuesday." It refers to Shrove Tuesday, the day before the Christian season of Lent begins. Mardi Gras is a time of great celebration, with people wearing colorful masks and costumes parading through the streets of New Orleans.

Louisiana's capital and second-largest city is Baton Rouge. It lies northwest of New Orleans on the Mississippi River.

For more information about Louisiana call (800) 227-4386

Top, boiled crawfish; *clockwise from lower left,* Honey Island Swamp; the capitol in Baton Rouge; New Orleans architecture; Jungle Gardens Bird Sanctuary on Avery Island; New Orleans jazz scene.

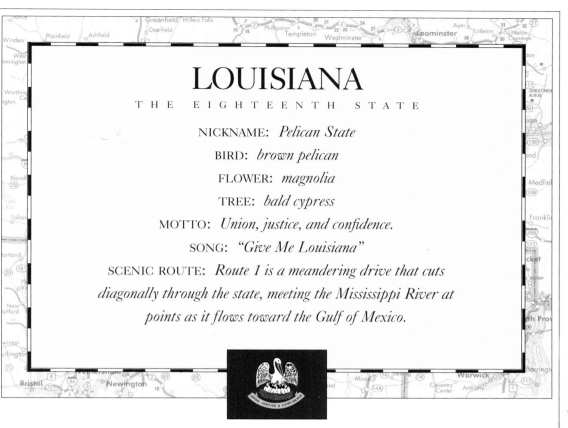

LOUISIANA
THE EIGHTEENTH STATE

NICKNAME: *Pelican State*

BIRD: *brown pelican*

FLOWER: *magnolia*

TREE: *bald cypress*

MOTTO: *Union, justice, and confidence.*

SONG: *"Give Me Louisiana"*

SCENIC ROUTE: *Route 1 is a meandering drive that cuts diagonally through the state, meeting the Mississippi River at points as it flows toward the Gulf of Mexico.*

MAINE

THE TWENTY-THIRD STATE

NICKNAME: *Pine Tree State*

BIRD: *chickadee*

FLOWER: *white pine cone and tassel*

TREE: *white pine*

MOTTO: *I direct.*

SONG: *"State of Maine Song"*

SCENIC ROUTE: *Route 1 skirts the coast, providing breathtaking views of the Atlantic Ocean and the state's many bays.*

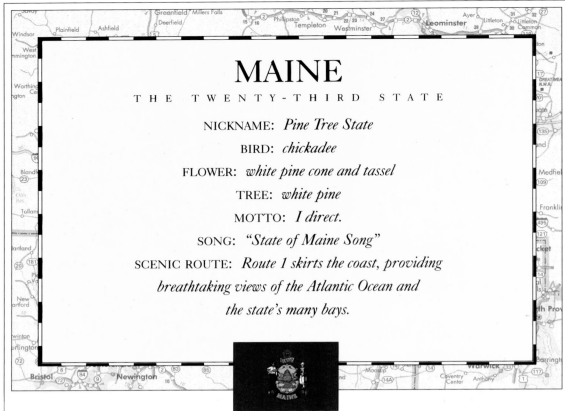

MAINE is the northernmost state in New England and extends farther east than any other point in the country. Those who climb to the top of Maine's mile-high Mount Katahdin at dawn could be the first to see the sun rise over the United States. The famed Appalachian Trail, which winds through 14 states, begins there in Maine.

Maine is known as the "Pine Tree State." More than 80 percent of its land is covered by forests of leafy and evergreen trees. Lumberjacks cut down selected trees and take the logs to lumber mills where they are sawed into boards. Leftover wood pulp is made into paper. Other workers use these supplies to make such things as furniture, canoes, baseball bats, and books. No other state makes more toothpicks than Maine, which produces almost 100 million each day.

With hundreds of offshore islands and a coastline jagged with bays and rocky coves, Maine is also known for its fishing industry. More lobsters, about 22 million pounds (10 million kilograms), are caught in Maine than in any other state. Clams are plentiful along the Atlantic shore. The largest island off the Maine coast is Mount Desert Island, which is part of the beautiful Acadia National Park. The name *Maine* may have meant "mainland." Early explorers often called the mainland "The Main" so it would not be confused with the islands offshore.

In June 1775, the first naval battle of the Revolutionary War took place off Maine's coast. It was a success for the patriots, who captured the armed British ship, *Margaretta*, not far from Machias. Fort Western, in the capital city of Augusta, is where Benedict Arnold and his troops met to begin a march to attack British troops at Quebec that same year. The fort is open to visitors.

Portland is Maine's largest city and a key trading port. The boyhood home of the poet Henry Wadsworth Longfellow is found there. Former U.S. president George Bush has a family home in Kennebunkport, located farther south along the coast.

For more information about Maine call (800) 533-9595

Clockwise from lower left, Mt. Katahdin and Compass Pond; Bass Harbor Lighthouse, Acadia National Park; Rockport lobsters; *top,* lobster boat in Rockland Harbor.

MARYLAND ranks forty-second in size among the 50 states, but is nineteenth in number of people. The Chesapeake Bay splits Maryland into two shores. Ships from around the world travel up this mighty waterway to the busy port of Baltimore, Maryland's largest city. It was there, on September 13, 1814, that Francis Scott Key watched British ships attack Fort McHenry. At dawn the next morning, his joy at seeing the American flag still flying above the fort inspired him to pen the words to "The Star-Spangled Banner," the national anthem of the United States.

Visitors to Baltimore today can tour Fort McHenry or see the National Aquarium and the Maryland Science Center in the Inner Harbor. The U.S.S. *Constellation*, a 36-gun frigate built in 1797, is docked there. It was the first American warship to capture a foreign vessel.

South of Baltimore and along the Chesapeake Bay is Annapolis, Maryland's capital. The United States Naval Academy is located there.

The waters and surrounding shores of the Chesapeake Bay are a haven for wild ducks and Canada geese. Maryland leads the nation in oyster production and is second in producing blue crabs. Many people vacation on the sandy beaches of Ocean City on Maryland's Eastern Shore. The famous Chincoteague ponies graze in the marshy meadows of nearby Assateague Island National Seashore Park. With its ocean coast, large bay, and tidal rivers, Maryland has 3,190 miles (5,134 kilometers) of coastline.

In the western part of the state, people enjoy hiking in the Allegheny Mountains or camping at Deep Creek Lake. The presidential retreat, Camp David, is located on Catoctin Mountain near Thurmont. Antietam National Battlefield near Sharpsburg marks the site of the bloodiest day of fighting in the Civil War. On September 17, 1862, over 10,000 Confederate and 12,000 Union soldiers were killed or wounded there.

Maryland was named after England's Queen Henrietta Maria, wife of King Charles I. It is known as the "Old Line State." In the Revolutionary War, a brave line of Maryland troops held back the British Army so that George Washington and his troops could escape capture. By doing so, 308 of the 404 Maryland fighters died.

For more information about Maryland call (800) 543-1036

Top, Swallow Falls State Park; *clockwise from lower left*, horse farm in Hunt Valley, north of Baltimore; aerial view of Baltimore's Inner Harbor; the U.S. Naval Academy, Annapolis.

MARYLAND
THE SEVENTH STATE

NICKNAME: *Old Line State*

BIRD: *Baltimore oriole*

FLOWER: *black-eyed Susan*

TREE: *white oak*

MOTTO: *Manly deeds, womanly words.*

SONG: *"Maryland, My Maryland"*

SCENIC ROUTE: *Route 219, in western Maryland, winds through Deep Creek Lake State Park, a popular location for outdoor activities throughout the year.*

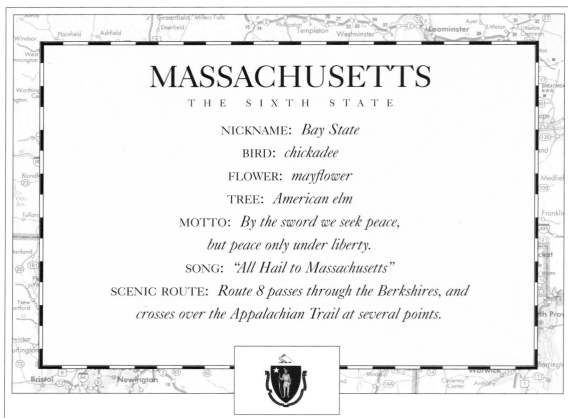

MASSACHUSETTS
THE SIXTH STATE

NICKNAME: *Bay State*

BIRD: *chickadee*

FLOWER: *mayflower*

TREE: *American elm*

MOTTO: *By the sword we seek peace,*
but peace only under liberty.

SONG: *"All Hail to Massachusetts"*

SCENIC ROUTE: *Route 8 passes through the Berkshires, and*
crosses over the Appalachian Trail at several points.

MASSACHUSETTS is named after the Massachusett Indian tribe, who lived in the area when the Pilgrims arrived. Visitors to Plimoth Plantation can see a recreation of the first Pilgrim village and an Indian settlement. They can also tour the *Mayflower II*, a copy of the Pilgrims' original ship, at dock in Plymouth.

Boston is the capital and largest city of Massachusetts. Nearly half of the population lives in or around the Boston area. Other large cities include Worcester and Springfield.

Throughout history, Massachusetts and its people have continued to play important roles. Harvard, the first college in the colonies, was established in 1636 at Cambridge. The first shots of the Revolutionary War were fired at Lexington and Concord. In historic Boston, red lines on the sidewalks mark the "Freedom Trail," a path leading to Revolutionary War places such as Paul Revere's house and Old North Church. *Old Ironsides*, a famous battleship from the War of 1812, is anchored nearby in the Boston harbor.

Famous American writers such as Nathaniel Hawthorne, Louisa May Alcott, Henry David Thoreau, and Emily Dickinson were all born in the "Bay State." Several of their homes are open as museums today. Four presidents were born in Massachusetts: John Adams, John Quincy Adams, John F. Kennedy, and George Bush.

From Massachusetts' earliest days, fishing has been a prime industry. Many of the houses in the port of Gloucester in northeastern Massachusetts have roof walks where, long ago, the wives of fishermen would watch for their husbands' safe return from the sea. Massachusetts remains among the leading states in commercial fishing.

Cape Cod National Seashore juts out like a large fishhook into the Atlantic Ocean. It has woodlands, marshes, and long stretches of sandy beaches. About half of America's cranberries are grown in bogs along the shore. Martha's Vineyard and Nantucket are islands off the southeastern coast of Massachusetts. These are popular vacation areas. The Berkshire Hills in western Massachusetts is also a scenic area to visit.

For more information about Massachusetts call (800) 447-6277

Clockwise from lower left, Plimoth Plantation; Faneuil Hall Marketplace, Boston; red barn in Shutesbury; Old North Church spire and statue of Paul Revere, Boston; *top,* Nauset Lighthouse, Eastham; *opposite page,* North Truro, Cape Cod.

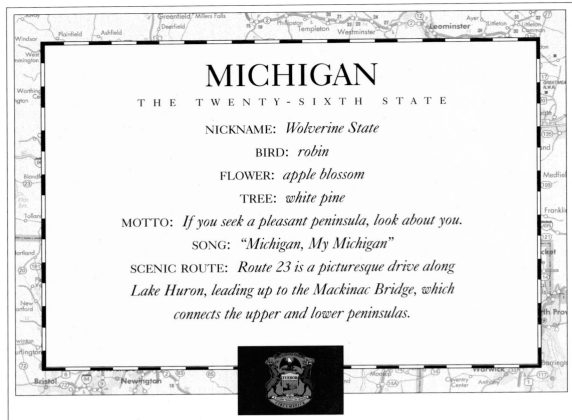

MICHIGAN

THE TWENTY-SIXTH STATE

NICKNAME: *Wolverine State*

BIRD: *robin*

FLOWER: *apple blossom*

TREE: *white pine*

MOTTO: *If you seek a pleasant peninsula, look about you.*

SONG: *"Michigan, My Michigan"*

SCENIC ROUTE: *Route 23 is a picturesque drive along Lake Huron, leading up to the Mackinac Bridge, which connects the upper and lower peninsulas.*

ALTHOUGH it is an inland state, Michigan has 3,288 miles (5,292 kilometers) of shoreline, more than any state except Alaska. Michigan's two peninsulas touch four of the five Great Lakes. The name *Michigan* was taken from the Chippewa Indian word *Michigama*, meaning "great lake." There are 11,000 smaller lakes within the state.

Completed in 1957 for $100 million, the 5-mile (8-kilometer) Mackinac bridge, called the "Big Mac," links Michigan's Upper Peninsula and Lower Peninsula. On nearby Mackinac Island, visitors can tour Fort Mackinac, which was built by the British in 1780. No cars are allowed on this resort island.

Michigan's Upper Peninsula has large deposits of copper and iron as well as thick forests abundant with wildlife. Years ago, fur traders brought wolverine pelts to trading posts there, resulting in Michigan's nickname the "Wolverine State."

Tall cliffs and rocks form unusual shapes and designs at Pictured Rocks National Lakeshore, which is near Munising on Lake Superior. At the National Ski Hall of Fame in Ishpeming, exhibits of famous U.S. skiers and skiing events can be seen.

The Lower Peninsula is known for its excellent farmland and centers of manufacturing. Much of the cereal eaten in the United States is processed in Michigan. In fact, more breakfast cereal is produced in Battle Creek than in any other city in the world.

Michigan leads all states in making cars and trucks. Detroit, Michigan's largest city, is known as the "Motor City." There, in 1896, Henry Ford built his first workable automobiles. West of Detroit is another important city for automaking, Lansing, the state capital. Tours of the Ford Motor Company factory are given in nearby Dearborn. At Greenfield Village, visitors can ride in an old Model-T Ford automobile and see the relocated homes of Thomas Edison, the Wright brothers, and other great inventors.

For more information about Michigan call (800) 5432-YES (543-2937)

Clockwise from lower left, Knob Hill, Mackinac Island; Detroit waterfront; Mackinac Bridge; Miner's Castle, Pictured Rocks; *top,* Lake Michigan, Leland.

THE name *Minnesota* is taken from two Sioux Indian words meaning "sky-colored waters." The state has at least 15,000 lakes, along with several beautiful waterfalls. Minnesota regularly gives out more fishing licenses than any other state.

Northern Minnesota is heavily forested. At Lumbertown, U.S.A., in central Minnesota, visitors can tour a full-scale model of a lumber town from the 1870s. Tall tales claim that Minnesota's lakes were formed in the footprints made by Paul Bunyan's giant blue ox, Babe. Huge statues of the legendary giant lumberjack and his ox can be seen in Brainerd and Bemidji. Not far from Bemidji is Lake Itasca, where the Mississippi River begins. This largest river in the United States starts out as a small stream about 10 feet (3 meters) wide and 2 feet (61 centimeters) deep.

Minnesota's tremendous iron ore deposits make it the number one iron-mining state. At Hibbing, the open-pit mines are so big that they are known as the "Grand Canyons of the Mesabi Range." Much of the iron that is mined there is shipped to Cleveland and Pittsburgh to make steel.

Pipestone National Monument in southwestern Minnesota has a type of red stone not found anywhere else in the country. For centuries, Native Americans have come to these sacred grounds to make peace pipes from the red pipestone found there. In southeastern Minnesota is the city of Rochester, where one of the world's greatest medical centers, the Mayo Clinic, is located. The southern part of the state is largely prairie. The number of gophers living there has prompted Minnesota's nickname of the "Gopher State."

Half of Minnesota's population lives in the twin cities of Minneapolis, the state's largest city, and St. Paul, the state's capital. Minnehaha Falls in Minneapolis is a beautiful waterfall made famous in Henry Wadsworth Longfellow's poem "The Song of Hiawatha." A bronze statue of Hiawatha and his Native American bride, Minnehaha, stands at the top of the falls. At nearby Fort Snelling, visitors can tour the restored military outpost built from 1820 to 1825 and learn more about life in the early nineteenth century.

For more information about Minnesota call (800) 657-3700

Top, Cascade River and Falls; *clockwise from lower left,* cross-country skiers on the Gunflint Trail; Native American dancer, Grand Portage Powwow; headwaters of the Mississippi River, Itasca State Park; Minneapolis Sculpture Garden; Minnesota farm scene.

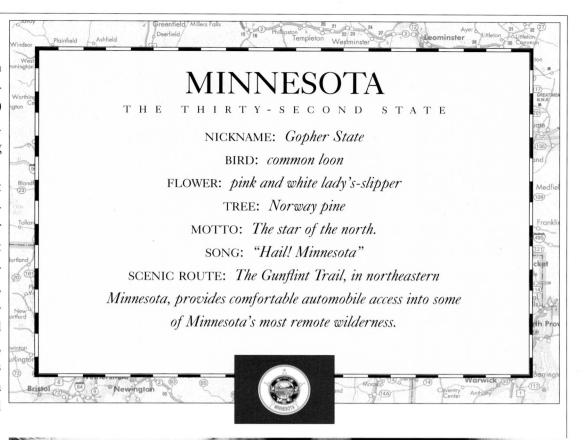

MINNESOTA
THE THIRTY-SECOND STATE

NICKNAME: *Gopher State*

BIRD: *common loon*

FLOWER: *pink and white lady's-slipper*

TREE: *Norway pine*

MOTTO: *The star of the north.*

SONG: *"Hail! Minnesota"*

SCENIC ROUTE: *The Gunflint Trail, in northeastern Minnesota, provides comfortable automobile access into some of Minnesota's most remote wilderness.*

MISSISSIPPI

THE TWENTIETH STATE

NICKNAME: *Magnolia State*

BIRD: *mockingbird*

FLOWER: *magnolia*

TREE: *magnolia*

MOTTO: *By valor and arms.*

SONG: *"Go, Mississippi"*

SCENIC ROUTE: *Natchez Trace Parkway is one of the most scenic roads in the U.S. It follows an old Native American Trail from Natchez to Nashville, Tennessee.*

MISSISSIPPI takes its name from the great river that forms 400 miles (644 kilometers) of its western boundary. *Mississippi* is an Indian word meaning "father of waters." The Mississippi River is the largest river in the United States and an important pathway into the heart of the country.

Much of Mississippi life is influenced by the waters of the Mississippi River. Strong currents and high water levels have carried rich topsoil from upriver to Mississippi's shores. In this fertile farmland, cotton and soybeans are grown. To prevent frequent flooding, levees were built along the riverbanks to hold back rising waters.

Southeastern Mississippi borders on the Gulf of Mexico, where year-round warm temperatures and large sandy beaches attract many visitors. Fishing is big business in Mississippi, a leader in catching shrimp and oysters. The seaport cities of Pascagoula and Biloxi are major centers for shipbuilding. All types of ships, from merchant vessels to atomic submarines, are made there.

On nearby Ship Island, visitors can tour Fort Massachusetts. It was used as a prison by the Union Army during the Civil War. Another important Civil War site is at Vicksburg National Military Park. After 47 days of fighting from May to July 1863, Union troops finally captured the Confederate stronghold in Vicksburg. This victory enabled the Union Army to control the Mississippi River.

In southwestern Mississippi is Natchez, a city that was twice designated as the state's capital (1798–1802, 1817–21). The Natchez region has many beautiful old mansions that stand today as reminders of the pre–Civil War South. Jackson has been the state's capital since 1822, and is also Mississippi's largest city.

Magnolia trees grow throughout the state. It was from these evergreen trees with white blossoms that Mississippi got its nickname of the "Magnolia State."

For more information about Mississippi call (800) 927-6378

Clockwise from lower left, Vicksburg National Military Park; Cypress Swamp, Natchez Trace Parkway; Ship Island; Mississippi Delta cotton fields; *top,* pilgrimage tours.

MISSOURI is one of two states whose borders touch eight states. (The other is Tennessee.) Missouri was named after the Missouri River, which flows from west to east across the state. The word *Missouri* is from an Indian word that means "town of the large canoes." The mighty Mississippi River and the Missouri River partly frame the state. Once, Native Americans paddled these waterways. Then steamboats became the most popular mode of transportation. Today, large barges haul wheat and other products from Kansas City across to St. Louis, the two largest cities in the state. From St. Louis, the barges head down the Mississippi River. St. Louis was the state capital until 1826; since then Jefferson City, in the central part of the state, has been capital.

In the mid-1800s, thousands of pioneers searching for new homes traveled to Missouri, where they could more easily head west on the Oregon and Santa Fe trails. Today, St. Louis' giant Gateway Arch serves as a reminder of Missouri's key position in history as the "Gateway to the West." The rainbow-shaped monument reaches a height of 630 feet (192 meters). Visitors ride in elevator cars that move like a Ferris wheel to an observation deck at the top of the arch. From there, they can see a wide expanse of cities, prairies, and rivers.

Farther north on the Mississippi River is the town of Hannibal. This was the boyhood home of the famous writer Samuel Clemens, better known by his pen name of Mark Twain. As a child, Clemens played on the islands and in the caves along the Mississippi River. Later, in such books as *The Adventures of Huckleberry Finn* and *The Adventures of Tom Sawyer,* his childhood experiences were a source of inspiration. The Mark Twain Home and Museum in Hannibal are open to the public.

The beautiful Ozark Mountains are in southern Missouri. These rugged hills are popular vacation spots for campers, swimmers, and fishing enthusiasts. More than 1,450 caves are found there.

The "Show Me State" is Missouri's nickname. It is said that people from Missouri have to see something to believe it, but the phrase actually came from a speech given by a Missouri congressman in 1899. "I am from Missouri," he said. "You have got to show me."

For more information about Missouri call (800) 877-1234

Top, the Gateway Arch, St. Louis; clockwise from lower left, Alley Spring Mill, Eminence; Taum Sauk Mountain, Ironton; lock and dam, Mississippi River; Mark Twain House, Hannibal.

MISSOURI

THE TWENTY-FOURTH STATE

NICKNAME: *Show Me State*

BIRD: *bluebird*

FLOWER: *hawthorn*

TREE: *flowering dogwood*

MOTTO: *The welfare of the people shall be the supreme law.*

SONG: *"Missouri Waltz"*

SCENIC ROUTE: *Route 79 runs along the Mississippi River, on the border of Missouri and Illinois. It joins St. Louis and Hannibal, home of Samuel Clemens (Mark Twain).*

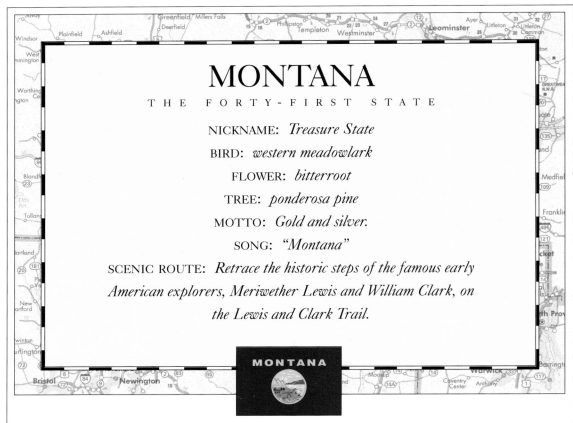

MONTANA
THE FORTY-FIRST STATE

NICKNAME: *Treasure State*

BIRD: *western meadowlark*

FLOWER: *bitterroot*

TREE: *ponderosa pine*

MOTTO: *Gold and silver.*

SONG: *"Montana"*

SCENIC ROUTE: *Retrace the historic steps of the famous early American explorers, Meriwether Lewis and William Clark, on the Lewis and Clark Trail.*

MONTANA

ONE of the nicknames for Montana is "Big Sky Country." A vast blanket of blue seems to go on forever above the cattle ranches and wheat fields on the state's eastern great plains. Montana is the fourth-largest state in size, yet it ranks forty-fourth in population. The few cities of any size, such as Billings and Great Falls, lie miles apart from each other. Some ranch children still go to school in one-room schoolhouses or move into cities during the school year to attend larger schools.

The Rocky Mountains cut through the western part of the state. The name *Montana* is from a Spanish word meaning "mountainous." Montana is called the "Treasure State" because of its rich mineral resources. Large deposits of copper, coal, oil, and many precious and semi-precious stones are found there, making it a rock collector's paradise.

In 1862, the discovery of gold brought thousands of settlers to the rugged hills of western Montana. Virginia City, in southwestern Montana, sprang up almost overnight as a gold camp. Today the town is rebuilt to look as it did during the gold rush. Helena, the state's capital, also started out as a mining camp.

At the Custer Battlefield National Monument, visitors can see where Lieutenant Colonel George Custer and about 200 cavalry soldiers attacked more than 2,000 Sioux and Cheyenne Indians led by Chief Crazy Horse. The battle took place at the Little Bighorn River on June 25, 1876. A cemetery now marks the place known as "Custer's Last Stand," where all of the cavalry soldiers—and Custer—lost their lives.

Montana has great stretches of unspoiled wilderness and an abundance of wildlife. Deer and pronghorn antelope live on the plains. Moose, elk, grizzly bear, mountain goat, and sheep are found in the mountains. Glacier National Park in northwestern Montana is a beautiful region of mountains, glaciers, and 250 lakes. Several of its peaks have not yet been climbed.

For more information about Montana call (800) VISIT MT (847-4868)

Left, a bald eagle soars through the Montana wilderness; *far left,* Little Bighorn Battlefield; *top left;* Autumn in Glacier National Park; *top,* Crow Fair, Hardin.

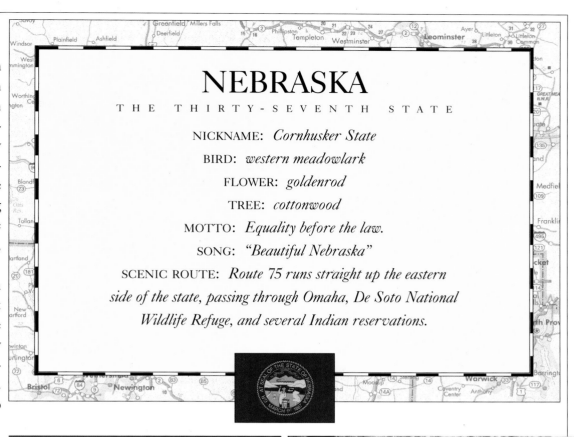

OVER 90 percent of the land in Nebraska is ranches and farms. Miles of corn grow in the fertile, black soil of eastern Nebraska. Years ago, farming families held cornhusking bees, where people socialized as they worked to remove the outer coverings on corncobs by hand. Today, Nebraska is known as the "Cornhusker State." Nebraska is also among the leading producers of beef cattle in the United States. On the north-central plains, huge herds of livestock graze.

The name *Nebraska* was taken from *nebrathka*, an Oto Indian word meaning "flat water." It was the name the Indians gave to the very broad and shallow Platte River. In July 1804, explorer William Clark wrote in his journal: "This great river . . . spreads very wide . . . and does not rise more than 5 or 6 feet (152 to 183 centimeters)."

Hundreds of thousands of pioneers traveled west along the Platte and North Platte rivers on the famous Oregon and Mormon trails. Today, ruts formed by covered wagons more than a century ago can still be seen. Near Bayard in western Nebraska is Chimney Rock, a famous landmark of the Oregon Trail. When pioneers reached this 500-foot-tall (150-meter-tall) rock formation, they knew that the Rocky Mountains were just ahead.

Many pioneers liked what they saw of Nebraska and decided to settle there. Because trees were scarce, settlers often built their homes out of sod. They would cut the sod into hard blocks and then stack them to build their houses. The term "sodbusters" was often used to refer to these early settlers. They would pack loose dirt or mud in between the sod blocks to hold them tightly together. In spring, wildflowers often bloomed on the roofs of these houses, some of which are still standing.

The largest city in Nebraska is Omaha, located in the far eastern part of the state. Omaha is one of the most important centers for the trading and railway transporting of farm products in the country. It is also a world leader in the processing of meat products. The second-largest city in Nebraska is Lincoln, the capital. A popular attraction there is the University of Nebraska State Museum, which has among its exhibits the largest fossil ever found of a mammoth.

For more information about Nebraska call (800) 228-4307

NEBRASKA

THE THIRTY-SEVENTH STATE

NICKNAME: *Cornhusker State*

BIRD: *western meadowlark*

FLOWER: *goldenrod*

TREE: *cottonwood*

MOTTO: *Equality before the law.*

SONG: *"Beautiful Nebraska"*

SCENIC ROUTE: *Route 75 runs straight up the eastern side of the state, passing through Omaha, De Soto National Wildlife Refuge, and several Indian reservations.*

Top, Millet, Gage County; *clockwise from lower right,* Scotts Bluff National Monument; the Capitol Building in Lincoln; Indian Cave State Park.

NEVADA

THE THIRTY-SIXTH STATE

NICKNAME: *Silver State*

BIRD: *mountain bluebird*

FLOWER: *sagebrush*

TREE: *single-leaf pinon*

MOTTO: *All for our country.*

SONG: *"Home Means Nevada"*

SCENIC ROUTE: *The Pony Express route, which parallels today's U.S. 50, is a one-of-a-kind opportunity to experience perhaps the last outpost of the true American West.*

NEVADA was named for the Sierra Nevada Mountains, one of more than 30 mountain ranges in the state. *Nevada* is from a Spanish word meaning "snow-covered," referring to the higher mountain ranges that sometimes remain snow-covered all year.

Sagebrush, with its silver-green leaves and large clusters of small white flowers, can be found growing abundantly in Nevada's valleys and deserts. It is the official state flower and also gave rise to one of Nevada's nicknames, the "Sagebrush State."

Nevada is the driest state in the nation. Only about nine inches (23 centimeters) of rain fall there each year. Most of Nevada is part of the Great Basin, a huge desert area that stretches into six states. Hoover Dam, one of the biggest dams in the world, was built on the Nevada-Arizona border in 1936 to help provide water for crops. The dam holds back water to form Lake Mead, one of the largest manmade lakes in the United States.

Very few people lived in what is now Nevada until the mid-1800s. When rich deposits of silver and gold were discovered in 1859, people flocked to the region by the thousands. Mining towns sprang up practically overnight. The huge Comstock Lode in western Nevada yielded deposits of silver and gold worth more than a billion dollars. In 1864, the territory of Nevada became a state and was nicknamed the "Silver State."

Nevada is still a leader in the mining industry. The state leads the nation in the production of gold, mercury, and magnesite.

Nevada today is probably best known for its legalized gambling. It is one of the few places in the United States that allows gambling on cards, dice, and other games of chance. Huge numbers of tourists visit the casinos and nightclubs in the state's two largest cities, Las Vegas and Reno.

Also popular among tourists is Lake Tahoe, which lies not far west from Carson City, Nevada's capital. This resort area offers beautiful scenery and opportunities for water sports, camping, and skiing.

For more information about Nevada call (800) NEVADA-8 (638-2328)

Clockwise from lower left, skiing at Lake Tahoe; Virginia City; the Excalibur Hotel in Las Vegas; Reno at night; Cathedral Gorge, Panaca; *top,* Hoover Dam.

EARLY Indians called the rugged, snow-covered peaks in northern New Hampshire *waumbek methna,* which means "white rock." Today, these peaks are called the White Mountains. In the largest section, known as the Presidential Range, individual mountains are named after American presidents. At 6,288 feet (1,917 meters), Mount Washington is the highest peak in the northeast United States. It is known for its extreme weather conditions. On April 12, 1934, the strongest wind ever recorded on earth—231 miles per hour (372 kilometers per hour)—whipped through Mount Washington.

Forty-seven peaks in the White Mountains rise above 4,000 feet (1,219 meters). Climbers who scale all of them can join the Four Thousand Footer Club, which now has more than 2,000 members. The oldest to join was in his seventies, and the youngest was a six-year-old boy! At Franconia Notch in the White Mountain region is the Flume—a deep, narrow valley 800 feet (240 meters) long. Visitors to the Flume can see some breathtaking wilderness scenes.

Most of New Hampshire's residents live in the southeastern part of the state. Beaches along the Atlantic coast are popular vacation spots in the warmer months. At Strawbery Banke in Portsmouth, visitors can tour a restored colonial seaport with homes built in the late 1700s.

New Hampshire was named after the county of Hampshire in England. It was the first of the original 13 colonies to declare independence from England. It is nicknamed the "Granite State" because of the huge deposits of granite underneath its surface. More than 80 percent of New Hampshire is forested. In autumn, the trees' blazing colors are among the most brilliant in the country. Many writers, including poet Robert Frost, have been inspired by New Hampshire's beauty.

The two largest cities in New Hampshire are Manchester and Nashua. Concord, the capital and third largest city, is located in the south-central part of the state.

For more information about New Hampshire call (800) 386–4664

NEW HAMPSHIRE
THE NINTH STATE

NICKNAME: *Granite State*

BIRD: *purple finch*

FLOWER: *purple lilac*

TREE: *white birch*

MOTTO: *Live free or die.*

SONG: *"Old New Hampshire"*

SCENIC ROUTE: *Route 112 takes drivers along the 36-mile Kancamagus Scenic Byway.*

Top, Marlborough foliage; *right,* Marlow; *far right,* skiing on Cannon Mountain; *upper right,* Andover.

41

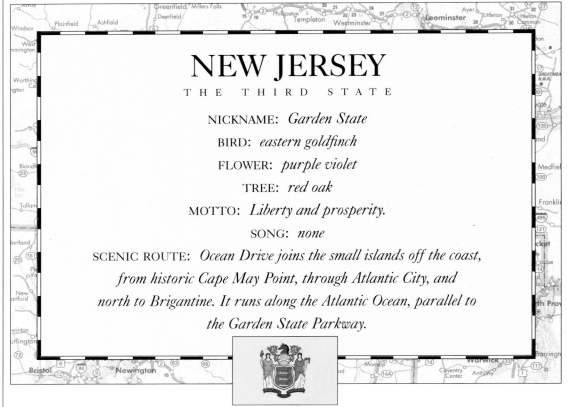

NEW JERSEY

THE THIRD STATE

NICKNAME: *Garden State*

BIRD: *eastern goldfinch*

FLOWER: *purple violet*

TREE: *red oak*

MOTTO: *Liberty and prosperity.*

SONG: *none*

SCENIC ROUTE: *Ocean Drive joins the small islands off the coast, from historic Cape May Point, through Atlantic City, and north to Brigantine. It runs along the Atlantic Ocean, parallel to the Garden State Parkway.*

NEW JERSEY was named after the island of Jersey in the English Channel. It is the most densely populated state. New Jersey has 1,050 people per square mile (407 people per square kilometer). Many people live in and around Newark, Jersey City, and Paterson, the three largest cities. The volume of traffic on the New Jersey Turnpike and the Garden State Parkway is among the heaviest in the United States.

Not all of the state is heavily populated. The southcentral area is rich farmland where many kinds of vegetables are grown. Fresh produce grown in the "Garden State" feeds people from Philadelphia to New York City. In southern New Jersey, 1.5 million acres (600,000 hectares) of pine trees and swamps make up the Pine Barrens. This wilderness area is an important ecosystem with unique species of plants and animals. Hiking, boating, swimming, and fishing are popular activities there.

America's most famous inventor, Thomas Edison, lived and worked in New Jersey. He is best known for inventing the electric light, the phonograph, and motion pictures. Visitors can walk through his research laboratory and home in West Orange.

Nearly 100 Revolutionary War battles were fought in New Jersey. One of General George Washington's best-known victories occurred when he and 2,400 soldiers braved a storm and crossed the icy Delaware River on Christmas night, 1776. They then marched to Trenton, now the state's capital, and defeated the surprised Hessian troops there.

Except for its small boundary with New York, New Jersey is surrounded by water. The Delaware River forms its western border, and the Hudson River separates it from New York in the east. Each year, millions of visitors travel to Atlantic City to swim in the ocean and walk on the world's most famous boardwalk, a 5-mile (8-kilometer) stretch of shops, hotels, and casinos. Atlantic City's streets bear names such as Park Place and Ventnor Avenue. Why do they sound familiar? The streets in the game of Monopoly were named after those in Atlantic City.

For more information about New Jersey call (800) JERSEY7 (537-7397)

Clockwise from lower left, Westfield; Atlantic City at night; camping in the Pine Barrens; Lamington; *top,* building sandcastles in Atlantic City.

THE beautiful scenery, sunny climate, and interesting Indian and Spanish cultures of New Mexico have earned it the nickname "Land of Enchantment." In no other state are the contrasts of yesterday and today so artfully blended. Gracious old Spanish missions and ancient pueblos stand comfortably near modern American structures. Santa Fe, the state's capital and second-largest city, is the oldest seat of government in the United States. In 1609-10, it became the capital of a Spanish province. The Palace of the Governors, built during the same period, still stands today and is the oldest government building on U.S. soil.

Farther north and east of Santa Fe is Taos Pueblo, one of the largest and best-known adobe villages built by Native Americans. Descendants of those Pueblo Indians still live in the buildings made more than 700 years ago. New Mexico has the country's fourth-largest Native American population.

New Mexico is the fifth-largest state in size. Flat plains, steep mountains, rugged canyons, and colorful deserts can be found within its borders. Much of the land is unspoiled wilderness.

Turquoise, copper, silver, gold, and many semiprecious stones are among New Mexico's chief mineral products. The state also has rich supplies of uranium, a main source of atomic energy. On July 16, 1945, the first atomic bomb was tested in a remote desert area near Alamogordo in south-central New Mexico. Today, scientists at Los Alamos in north-central New Mexico continue to experiment with other uses for nuclear energy.

Carlsbad Caverns, in southeastern New Mexico, are the largest known caves in the world. One cavern, known as "Big Room," is three-quarters of a mile (1.2 kilometers) long, 625 feet (190 meters) wide, and 300 feet (91 meters) high. Unusual rock formations, formed over thousands of years, are found throughout the caves. Each day, thousands of bats fly out of the caverns in the evening and return at dawn.

The largest city in New Mexico is Albuquerque. It was founded by the Spanish in 1706 and is a thriving industrial center.

For more information about New Mexico call (800) 545-2040

NEW MEXICO
THE FORTY-SEVENTH STATE

NICKNAME: *Land of Enchantment*

BIRD: *roadrunner*

FLOWER: *yucca*

TREE: *piñon*

MOTTO: *It grows as it goes.*

SONG: *"O, Fair New Mexico"*

SCENIC ROUTE: *Sunspot Highway, a 15-mile stretch of Route 6563, winds along the front rim of the Sacramento Mountains. White Sands National Monument is visible from the road.*

Top, Kiowa National Grasslands; *clockwise from lower left,* claret cup cactus, City of Rocks State Park; Balloon Fiesta, Albuquerque; Church of Saint Francisco of Asia, Taos; White Sands National Monument.

NEW YORK was named after England's Duke of York. It is a state of mountains, seashores, big cities, and wilderness. Camping and fishing are popular in the scenic Finger Lakes region.

The Adirondack Mountains, in northeastern New York, are well-situated for winter and summer sports. The 1932 and 1980 Winter Olympics were held at beautiful Lake Placid. Nearby Mount Marcy, 5,344 feet (1,629 meters) tall, is the highest peak in the state. In southeastern New York are the Catskill Mountains, where the title character of Washington Irving's story "Rip Van Winkle" slept for 20 years. Albany, the state capital, is nearby.

Many people travel to Cooperstown to visit the National Baseball Hall of Fame and Museum. Exhibits honoring baseball's best, as well as displays of baseball cards and equipment, are featured there.

Niagara Falls is one of the most spectacular sights in North America. It consists of two separate falls, the Horseshoe Falls on the Canadian side of the border and the American Falls in New York. On the New York side, about 5,664,000 cubic feet (169,920 cubic meters) of water go over the falls each minute. Several steamers, each called *The Maid of the Mist*, carry sightseers close to the thundering waters at the base of the falls.

In 1626, the Manhattan Indians sold the island of Manhattan to Dutch settlers for tools and beads worth $24. Today, Manhattan is part of New York City. Skyscrapers stand where Indian huts once stood. New York City is a major port, a leading center of business and culture, and one of the largest cities in the country. More than seven million people live there, more people than in 41 of the 50 states. Visitors can do many things, including taking a boat ride to Liberty Island to see the Statue of Liberty, or attending a play in a Broadway theater. New York City's Metropolitan Museum of Art is the country's largest art museum, and the New York Public Library is the largest public library of any U.S. city.

George Washington once predicted that New York would be the seat of a new empire. Perhaps that is the origin of the state's nickname, "Empire State."

For more information about New York call (800) CALL NYS (225-5697)

Opposite page, R.H. Treman Park, Ithaca; *top,* Statue of Liberty; *right,* West Point cadets on parade; *far right,* Niagara Falls; *top right,* South Street Seaport, Manhattan.

NEW YORK
THE ELEVENTH STATE

NICKNAME: *Empire State*

BIRD: *bluebird*

FLOWER: *rose*

TREE: *sugar maple*

MOTTO: *Ever upward.*

SONG: *"I Love New York"*

SCENIC ROUTE: *The Seaway Trail is a scenic and historic motor and waterway route along the St. Lawrence River and Lake Ontario.*

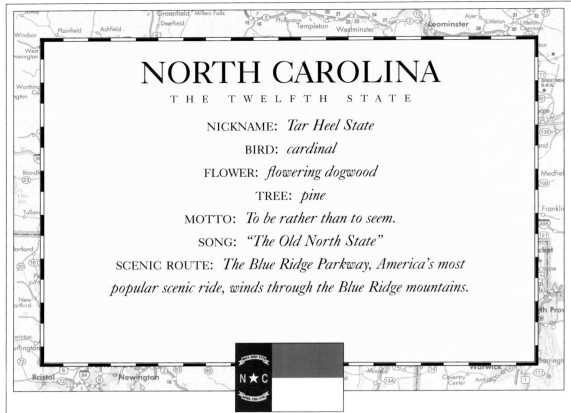

NORTH CAROLINA

THE TWELFTH STATE

NICKNAME: *Tar Heel State*

BIRD: *cardinal*

FLOWER: *flowering dogwood*

TREE: *pine*

MOTTO: *To be rather than to seem.*

SONG: *"The Old North State"*

SCENIC ROUTE: *The Blue Ridge Parkway, America's most popular scenic ride, winds through the Blue Ridge mountains.*

NORTH CAROLINA stretches about 500 miles (805 kilometers) from its Atlantic shoreline in the east to the Great Smoky Mountains National Park in the west. Its largest city is Charlotte, the nation's leader in the wholesale textile trade. Raleigh, the state capital, and Greensboro are other cities with sizable populations.

In the western part of the state, near Linville, is Grandfather Mountain, perhaps the oldest mountain in North America. Near the town of Lake Lure in western North Carolina stands Chimney Rock, 225 feet (69 meters) high. Visitors can drive, hike, or take an elevator to the top of it. Chimney Rock offers a beautiful view of the Blue Ridge Mountains. The Blue Ridge Parkway, America's most popular scenic ride, snakes along mountaintops to offer other spectacular views from many overlooks.

North Carolina is tobacco country. It is the leading tobacco-growing state in the nation. It is also number one in the manufacture of furniture and textiles.

Visitors to Kill Devil Hills, near Kitty Hawk in far eastern North Carolina, can see where Orville and Wilbur Wright made the first powered airplane flight on December 17, 1903. Farther down the coast is Cape Hatteras. It was once known as the "Graveyard of the Atlantic" because of all the ships that had wrecked there. Today, a 208-foot (63-meter) lighthouse, the tallest in the United States, stands there to warn ships of danger. Farther south of Cape Hatteras is Ocracoke Island, where the infamous pirate Blackbeard was killed.

The original American colony of Carolina was a large piece of land that extended from the Atlantic to the Pacific Ocean and included what is now North and South Carolina. It was named for King Charles I of England and was called the *Province of Carolana* ("land of Charles"). In 1663, the spelling was changed to *Carolina*. North Carolina and South Carolina became separate colonies in 1730.

North Carolina is known as the "Tar Heel State." Some people claim that the nickname was first used during the Civil War. It was said that soldiers from North Carolina stood firm in battle—as if tar, an early state product, kept their heels from retreating.

For more information about North Carolina call (800) VISIT NC (847–4862)

Clockwise from lower left, bagpipers on Grandfather Mountain; Charlotte; a Cherokee canoe carver; the Blue Ridge Parkway; Merchant's Mill Pond State Park; Wright Brothers National Memorial, Kill Devil Hills; *top,* Cape Hatteras lighthouse.

IN North Dakota, farms and ranches stretch from the scenic Badlands in the west to Red River Valley in the east. North Dakota is the number one state for producing barley and sunflower seeds. It is also a leading producer of wheat and rye. Like South Dakota, North Dakota was named after the Lakota, or Sioux, Indians who once hunted buffalo on the plains of the Midwest. North Dakota is known as the "Flickertail State," referring to the many flickertail ground squirrels found there. The geographic center of the entire North American continent is located near the town of Rugby in north-central North Dakota.

Today, North Dakota has four Indian reservations. Each summer, special Native American ceremonies are open to the public. Near Fort Yates on the Standing Rock Indian Reservation, the grave of the great Sioux leader Sitting Bull can be seen. Fort Abraham Lincoln State Park is restored to look as it did in 1876, the year Lieutenant Colonel George Custer and his troops rode out from the fort on a journey that would end at Montana's Little Bighorn, scene of their disastrous battle. Bonanzaville U.S.A., in West Fargo, is a preserved pioneer town of the 1800s. Nearby Slant Village has reconstructed lodges of the Mandan Indians.

When the Garrison Dam was completed across the Missouri River in 1960, Lake Sakakawea formed above it. It is 178 miles (286 kilometers) long and among the nation's largest manmade lakes. Far to the north of Lake Sakakawea is the International Peace Garden. This garden of beautiful flowers lies on both sides of the border between North Dakota and Manitoba, Canada, representing the friendship between the two countries.

Nearly four out of every ten North Dakotans live in or near the state's four largest cities – Fargo, Bismarck (state capital), Grand Forks, and Minot. These are the only cities with populations greater than 25,000.

For more information about North Dakota call (800) 435-5663

NORTH DAKOTA
THE THIRTY-NINTH STATE

NICKNAME: *Flickertail State*

BIRD: *western meadowlark*

FLOWER: *wild prairie rose*

TREE: *American elm*

MOTTO: *Liberty and union, now and forever, one and inseparable.*

SONG: *"North Dakota Hymn"*

SCENIC ROUTE: *Route 22 is a picturesque drive through Fort Berthold Indian Reservation and north to the top of Little Missouri National Grassland.*

Top, Sioux tepee, Anishinaubag Center on Fish Lake near Belcourt; *clockwise from lower left,* aerial view of a farm in the Red River Valley; International Peace Garden; waterfowl migrating through North Dakota along the Central Flyway; Bonanzaville USA; Theodore Roosevelt National Park, in the Badlands.

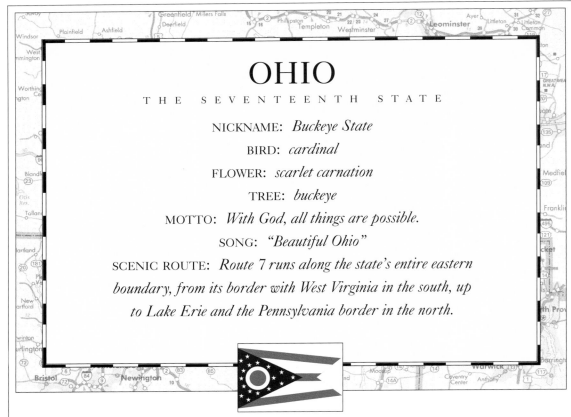

OHIO

THE SEVENTEENTH STATE

NICKNAME: *Buckeye State*

BIRD: *cardinal*

FLOWER: *scarlet carnation*

TREE: *buckeye*

MOTTO: *With God, all things are possible.*

SONG: *"Beautiful Ohio"*

SCENIC ROUTE: *Route 7 runs along the state's entire eastern boundary, from its border with West Virginia in the south, up to Lake Erie and the Pennsylvania border in the north.*

YEARS ago, the Iroquois Indians named a large river Ohio, meaning "something great." Today, the river is still called Ohio—as is the state above it. A major tributary of the Mississippi, the Ohio River carries more than 80 million tons of cargo each year—almost twice the amount that goes through the Panama Canal.

Ohio is nicknamed the "Buckeye State" because, at one time, buckeye trees were plentiful in the state. Many early settlers in Ohio used the wood of the buckeye tree to build their homes.

Since its early days of settlement, Ohio has been an important fruit producer. Grapes, strawberries, and apples are grown there. John Chapman, better known as Johnny Appleseed, spent most of his life planting apple trees in the Ohio Valley. He did this in the late eighteenth and well into the nineteenth century. A monument to him was built in the northern Ohio city of Ashland.

Ohio leads the nation in making rubber products. These include tires, hoses, and rubber bands. In the early 1900s, the northeastern city of Akron was the world leader in producing rubber tires. Only Indiana is ahead of Ohio in the production of iron and steel in the United States today.

The Great Serpent Mound, near Hillsboro in southern Ohio, was created by prehistoric people. It is formed out of earth and has seven deep curves. Built for burial or worship, it is more than a quarter of a mile (0.4 kilometer) long and is one of the best examples of ancient Indian earthworks that can be seen today.

Seven presidents were born in Ohio: Ulysses S. Grant, Rutherford B. Hayes, James A. Garfield, Benjamin Harrison, William McKinley, William Howard Taft, and Warren G. Harding. Because of this, Ohio is sometimes called the "Mother of Presidents," a nickname it shares with Virginia.

Ohio's four largest cities are Cleveland, Columbus, Cincinnati, and Toledo. Columbus, the state capital, is a center for scientific and technological information. Because its residents comprise a demographic cross-section of the United States, the city is often used for testing new products and menus, earning it the nickname "Test Market, U.S.A."

For more information about Ohio call (800) BUCKEYE (282-5393)

Left, Serpent Mound, Locust Grove; *far left,* Hocking Hills; *middle left,* Cincinnati at night; *top left,* the Valley Gem sternwheeler on the Ohio River, Marietta; *top,* scenic Athens.

IN the nineteenth century, the U.S. government often forced freed Native American families to move from the lands where they had lived for hundreds of years. Many of the families were sent to an area in the Midwest that was set aside for Native Americans. When that Native American territory became a state in 1907, it was given the name Oklahoma, which comes from two Choctaw Indian words meaning "red people."

Today, more Native Americans live in Oklahoma than in any other state. As many as 60 different tribes are represented. To see how early Indians lived, visitors can tour the Tsa-La-Gi Indian Village near Tahlequah in eastern Oklahoma. At this restored Cherokee village, Native Americans demonstrate crafts and talk about Indian traditions and customs.

Oklahoma has flat plains, mountains, deserts, and forests. Hidden beneath these grounds is a rich supply of oil, sometimes called "black gold." Oil wells can be seen throughout the state—there are even oil wells on the front lawn of the state capitol. The northwestern part, or Panhandle, of Oklahoma is 34 miles (55 kilometers) wide and 166 miles (267 kilometers) long. East to west, the state's rise toward the Rocky Mountains is significant. Oklahoma slants from a low elevation of 287 feet on the Little River in the southeast to a high of 4,973 feet at Black Mesa in the northwest corner of the panhandle.

In the late 1800s, pioneers were eager to settle in Oklahoma. Some came in sooner than was legally allowed by the federal government, earning Oklahoma the nickname of the "Sooner State." Many of these pioneers became ranchers, and today, Oklahoma is second only to Texas in raising beef cattle.

Oklahoma City, located in the central part of the state, is the state's capital and largest city. It was formed in one afternoon when the unassigned areas of the Oklahoma Territory were opened for settlement. Ten thousand land claims around a railroad station site were made between noon and sundown on April 22, 1889! The next largest city is Tulsa, in northeastern Oklahoma. About half of the state's population lives in or around these two cities.

For more information about Oklahoma call (800) 652-6552

Top; American Indian Expo, Anadarko; *right,* Black Mesa State Park, near Boise City; *far right,* hangliding on Buffalo Mountain, near Talihina; *top right,* the state capitol in Oklahoma City.

OKLAHOMA
THE FORTY-SIXTH STATE

NICKNAME: *Sooner State*
BIRD: *scissor-tailed flycatcher*
FLOWER: *mistletoe*
TREE: *redbud*
MOTTO: *Labor conquers all things.*
SONG: *"Oklahoma!"*
SCENIC ROUTE: *The 54-mile long Talimena Scenic Drive, S.H. 1, follows the crest of Winding Stair and Rich mountains from just northeast of Talihina.*

OKLAHOMA

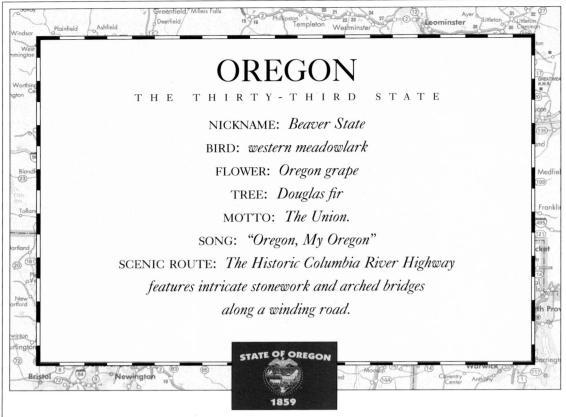

OREGON

THE THIRTY-THIRD STATE

NICKNAME: *Beaver State*

BIRD: *western meadowlark*

FLOWER: *Oregon grape*

TREE: *Douglas fir*

MOTTO: *The Union.*

SONG: *"Oregon, My Oregon"*

SCENIC ROUTE: *The Historic Columbia River Highway*
features intricate stonework and arched bridges
along a winding road.

STATE OF OREGON
1859

OREGON is the number one lumber-producing state. Nearly half of the land is covered with beautiful forests of evergreen trees. Some Douglas firs grow to heights of 250 feet (76 meters)—as tall as a 20-story building! Many Oregon factories manufacture a variety of wood products.

Elks, bobcats, wolves, and coyotes are among the animals that live in the forests of Oregon. Nineteenth-century fur trappers were drawn to this region in great numbers. Oregon is known as the "Beaver State" because of the thousands of beaver skins taken at that time.

The Cascade Mountains in Oregon have some of the highest peaks in North America. Mount Hood is the state's tallest peak at 11,239 feet (3,426 meters). South and west of Mount Hood is Mount Jefferson, 10,497 feet (3,199 meters) high. The Cascade Mountains also have the deepest lake in the United States, Crater Lake. Formed after the top of a volcano caved in, Crater Lake is 1,932 feet (589 meters) deep.

In the 1840s and 1850s, thousands of pioneers traveled across the country on the Oregon Trail, heading for the fertile soil of the Willamette Valley. Today, that area includes Oregon's largest city and seaport, Portland, and the capital city of Salem. Eugene is also an important metropolitan area.

The Columbia River flows west to the Pacific Ocean and forms most of the northern border of Oregon. Many people believe the name *Oregon* came from the French trappers' name for this river. They called it *ouragan,* meaning "hurricane." Today, many dams help control the massive water flow of the Columbia River. The Bonneville Dam in northwestern Oregon has fish ladders, a series of small stairstep dams built to help the Pacific salmon swim upstream to lay their eggs in the brooks where they were born.

Just north of Florence, along Oregon's Pacific coast, are the Sea Lion Caves. These are some of the largest sea caves in the world. Hundreds of sea lions live there between fall and summer, when they swim north to Alaska.

For more information about Oregon call (800) 547-7842

Left, Rim Trail, Crater Lake National Park; *far left,* Mount Hood; *top left,* Little Clearwater Creek Falls; *top,* a lumberjack scales a tree.

WILLIAM PENN founded Pennsylvania in 1681. King Charles II of England gave it to him in payment of a debt owed to Penn's father. William Penn named it *Sylvania*, a form of the Latin word meaning "woods." King Charles placed the word *Penn* in front of *Sylvania*, producing the final name of *Pennsylvania*.

Pennsylvania is known for its beautiful rivers and streams and many covered bridges. It is divided diagonally by several scenic mountain ranges and valleys. About 60 percent of Pennsylvania is forests.

East of Harrisburg, the state capital, is Hershey, a city chocolate lovers especially enjoy. It began as a planned community for the workers in Milton S. Hershey's chocolate plant. Today, visitors can tour the world's largest chocolate factory and watch as Hershey bars are being made.

Pennsylvania is nicknamed the "Keystone State." A keystone is the top stone in the middle of an arch and holds the other stones of the arch in place. In many ways, Pennsylvania helped hold the first states together. The two most important documents of the United States, the Declaration of Independence and the U.S. Constitution, were adopted in the Pennsylvania State House. Later renamed Independence Hall, this historic building is in downtown Philadelphia, the state's largest city and the fourth-largest city in the nation. The Liberty Bell, a symbol of America's freedom, stands nearby.

Gettysburg National Military Park in southern Pennsylvania was the scene of a bloody Civil War battle in July 1863. In three days of fighting, more than 38,000 soldiers were killed. The Union victory was a turning point in the Civil War. This is also the place where President Abraham Lincoln gave his most famous speech, the Gettysburg Address, in November of the same year.

Pennsylvania is one of the nation's top producers of coal and steel. About 10 percent of all the steel made in the United States annually is produced in and around Pittsburgh, the state's second-largest city,

For more information about Pennsylvania call (800) 847–4872

PENNSYLVANIA
THE SECOND STATE

NICKNAME: *Keystone State*
BIRD: *ruffed grouse*
FLOWER: *mountain laurel*
TREE: *hemlock*
MOTTO: *Virtue, liberty, and independence.*
SONG: *none*
SCENIC ROUTE: *Route 34 is a beautiful and historic drive through Gettysburg National Military Park and its surrounding areas.*

Top, the Liberty Bell, Philadelphia; clockwise from lower left, an Amish buggy, Lancaster; Independence Hall, Philadelphia; the Delaware Water Gap; Valley Forge.

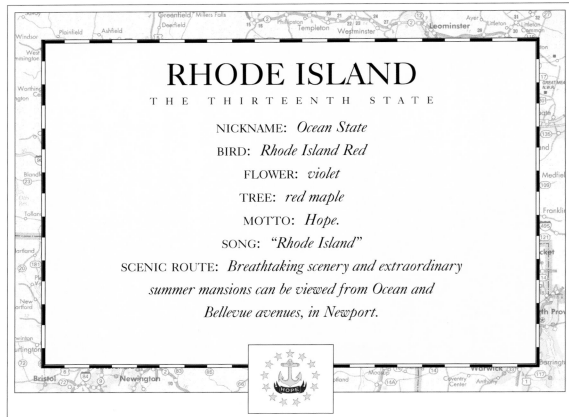

RHODE ISLAND

THE THIRTEENTH STATE

NICKNAME: *Ocean State*

BIRD: *Rhode Island Red*

FLOWER: *violet*

TREE: *red maple*

MOTTO: *Hope.*

SONG: *"Rhode Island"*

SCENIC ROUTE: *Breathtaking scenery and extraordinary summer mansions can be viewed from Ocean and Bellevue avenues, in Newport.*

RHODE ISLAND is the smallest state in size, but it has the longest official name—State of Rhode Island and Providence Plantations. It measures only 37 by 48 miles (60 by 77 kilometers). It would take 483 Rhode Islands to equal the size of Alaska.

Rhode Island, nicknamed the "Ocean State," has 384 miles (618 kilometers) of shoreline. Narragansett Bay slices into the eastern section of the state. Rhode Island also has 36 offshore islands with such interesting names as Hope, Patience, Hog, and Prudence. The largest of the islands is also named Rhode Island. Saltwater fishing is big business in the state.

Each year, the America's Cup yacht races are held in the waters off Newport. For years, many wealthy families have been attracted to Newport's mild climate and beautiful scenery. A number of the mansions built by those families can be seen from Cliff Walk. This is a 3-mile (5-kilometer) pathway along Newport's Atlantic coastline. Perhaps the most famous mansion to be seen there is "The Breakers." It has 70 rooms and was built in 1895 for Cornelius Vanderbilt, a powerful railroad businessman. The International Tennis Hall of Fame and Museum, also in Newport, features exhibits of tennis history and memorabilia of top tennis players.

In search of religious freedom, a minister named Roger Williams founded Rhode Island's first permanent European settlement at Providence in 1636. Encouraged by the colony's religious tolerance, many Jews settled in Rhode Island. The Touro Synagogue in Newport, founded in 1763, is believed to be the oldest in the nation. Over 60 percent of all Rhode Islanders live in or around Providence, the state capital and largest city. It is one of the world's largest centers for manufacturing jewelry and silverware.

No one really knows how Rhode Island got its name. Some people think it came from the Dutch words Roodt Eylandt, the name a Dutch sailor, Adriaen Block, gave to one of the offshore islands he saw in 1614. Roodt Eylandt means "Red Island," referring to the red clay Block saw on that island's shore.

For more information about Rhode Island call (800) 556-2484

Clockwise from lower left, sailing off Rhode Island's coast; Cliffwalk, Newport; the International Tennis Hall of Fame, Newport; the state capitol, Providence; *top,* Block Island lighthouse.

SOUTH CAROLINA is home to magnolia trees and Spanish moss, mountains and beaches, tobacco and palmetto trees. South Carolina's nickname, the "Palmetto State," probably came from the number of palmettos (small palm trees with fan-shaped leaves) that grow there.

Northwestern South Carolina is made up of the Piedmont and Blue Ridge areas, together called the "Up Country." In the Blue Ridge area is Sassafras Mountain, the highest peak in the state at 3,554 feet (1,083 meters). The central part of South Carolina is dotted with fields of tobacco, soybeans, peaches, and cotton. The state is second only to North Carolina in the production of textiles.

The eastern part of South Carolina's Atlantic coastal plain is called the "Low Country." Every year, thousands of people flock to the sandy beaches there, including Myrtle Beach and Hilton Head Island. From 1716 to 1718, Blackbeard and his band of pirates sailed and looted off this coastline.

In 1526, 500 Spaniards started a colony near Winyah Bay along the coast. Although the colony did not last, it was the first European settlement on land that is now the United States. The original American colony of Carolina was a huge piece of land that included what is now North and South Carolina and extended from the Atlantic to the Pacific Ocean. Named for King Charles I of England, it was called the *province of Carolana* (land of Charles). In 1663, the spelling was changed to *Carolina*. North and South Carolina became separate colonies in 1730.

More Revolutionary War battles (137 in all) were fought in South Carolina than in any other state. Along the coast, Francis Marion, the "Swamp Fox" hero, attacked British troops and then disappeared. Another leader of South Carolina troops during the Revolutionary War was Thomas Sumter. The fort named after him in Charleston was where the first shots of the Civil War were fired on April 12, 1861.

Columbia is South Carolina's capital and largest city. Other large cities include Charleston, North Charleston, Greenville, and Spartanburg.

For more information about South Carolina call (800) 346-3634

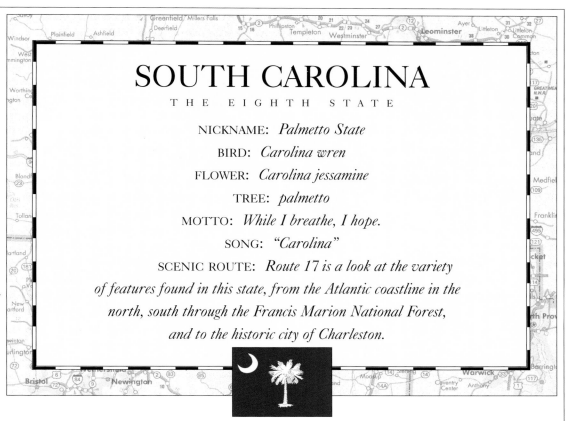

SOUTH CAROLINA
THE EIGHTH STATE

NICKNAME: *Palmetto State*
BIRD: *Carolina wren*
FLOWER: *Carolina jessamine*
TREE: *palmetto*
MOTTO: *While I breathe, I hope.*
SONG: *"Carolina"*
SCENIC ROUTE: *Route 17 is a look at the variety of features found in this state, from the Atlantic coastline in the north, south through the Francis Marion National Forest, and to the historic city of Charleston.*

Top, rafting on the Chattooga River; *clockwise from lower left,* Fort Sumter; Myrtle Beach; Hilton Head Island.

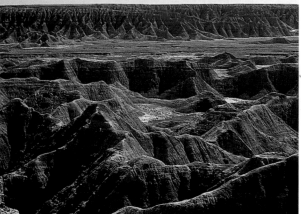

SOUTH DAKOTA
THE FORTIETH STATE

NICKNAME: *Sunshine State*

BIRD: *ring-necked pheasant*

FLOWER: *American pasqueflower*

TREE: *Black Hills spruce*

MOTTO: *Under God, the people rule.*

SONG: *"Hail, South Dakota"*

SCENIC ROUTE: *Needles Highway, a breathtaking 14-mile drive on SD 87, winds its way past towering granite spires, around hairpin curves, and through narrow tunnels.*

FOR hundreds of years, many Native Americans lived on the great plains that are now South Dakota. Some farmed, while others hunted huge roaming herds of buffalo. The Sioux Indian tribe called themselves *Lakota,* a word meaning "friends." South Dakota is named after these Sioux Indians. Its most popular nickname is the "Sunshine State" because of its sunny climate. It shares this nickname with Florida.

Many battles between Native Americans and the U.S. government were fought in South Dakota. In 1890, the massacre of Native Americans by troops at Wounded Knee in southwestern South Dakota was one of the last of these battles. The Native Americans were forced to give up most of their land. Today, there are eight Native American reservations in South Dakota. Only Arizona has more land set aside for reservations.

The slopes of South Dakota's beautiful Black Hills in the southwest are covered with evergreen trees. From a distance, they appear to be black in color. Harney Peak, at 7,242 feet (2,207 meters) is South Dakota's highest point.

The Homestake Mine in Lead and the town of Deadwood are also in the Black Hills. The Homestake Mine started up in 1876 and still produces gold—no other gold mine in the United States produces more. The wild gold-mining town of Deadwood also sprang up in 1876. Calamity Jane and Wild Bill Hickok are two of the famous people buried in Deadwood's Mount Moriah Cemetery.

One of the world's largest sculptures is the Mount Rushmore National Memorial near Rapid City. The monument is an amazing carving formed in a granite mountainside by Gutzon Borglum. It shows the faces and part of the heads of presidents George Washington, Thomas Jefferson, Theodore Roosevelt, and Abraham Lincoln. On nearby Thunderhead Mountain, an even larger memorial is being built to Sioux Indian Chief Crazy Horse.

South Dakota is a major farming state. It also supplies about 15 percent of all gold produced in the country. Only Nevada ranks ahead of South Dakota in gold production. The capital of the state is Pierre, and the largest city is Sioux Falls.

For more information about South Dakota call (800) 732-5682

Left, the Badlands; *far left,* Mount Rushmore; *top left,* the Corn Palace, Mitchell; *top,* Custer State Park.

54

TENNESSEE extends from the Blue Ridge region and Great Smoky Mountains in the east to the Mississippi River in the west. The misty Great Smoky Mountains National Park was once an area where Davy Crockett and Daniel Boone roamed. Davy Crockett's log cabin birthplace can be seen near Limestone in northeastern Tennessee.

Great numbers of Cherokee Indians lived in those mountains at one time, and a smaller number live there today. The name *Tennessee* was taken from the word *Tanasie*, which was the name of a Cherokee village. Today, Tennessee's mountain people are known for their beautiful crafts and fine fiddle music.

Tennessee is called the "Volunteer State." Throughout history, its people have always been quick to defend the state whenever needed. More than 400 Civil War battles were fought in Tennessee. At the base of Lookout Mountain in the southern part of the state is the Confederama. Inside, the 1863 Battle of Chattanooga is acted out by thousands of tiny motorized soldiers. The top of Lookout Mountain offers a view of seven different states.

In the 1930s, under the guidance of the Tennessee Valley Authority and the U.S. Army Corps of Engineers, a huge network of dams was built throughout the state. Once unnavigable waterways were deepened, and navigation locks were built. As a result, Tennessee enjoys improved commerce, flood control, inexpensive electricity, and enhanced recreational areas.

The soybeans, tobacco, and cotton grown in Tennessee each year make it one of the leading states in the nation for all three crops. Tennessee ranks first in the nation for production of zinc, and it is the only state in the South that mines copper.

The capital city of Nashville is second only to New York City as a center for recording music. Near Nashville is the Grand Ole Opry House, a world-famous home for country and western music. Memphis, in the southwestern corner of the state, is Tennessee's largest city. Graceland, once the home of singer Elvis Presley, is located in Memphis.

For more information about Tennessee call (615) 741-2158

TENNESSEE
THE SIXTEENTH STATE

NICKNAME: *Volunteer State*

BIRD: *mockingbird*

FLOWER: *iris*

TREE: *tulip poplar*

MOTTO: *Agriculture and commerce.*

SONGS: *"Tennessee Waltz" and "Rocky Top"*

SCENIC ROUTE: *Route 441, Newfound Gap Road, meanders through the Great Smoky Mountain National Park, and into North Carolina.*

Top, the Grand Ole Opry, Nashville; *right*, the Incline Railway, Chattanooga; *top right*, Chimney Tops, Gatlinburg.

55

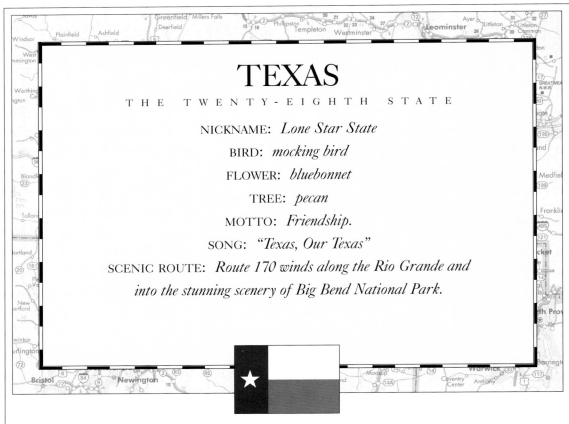

TEXAS

THE TWENTY-EIGHTH STATE

NICKNAME: *Lone Star State*

BIRD: *mocking bird*

FLOWER: *bluebonnet*

TREE: *pecan*

MOTTO: *Friendship.*

SONG: *"Texas, Our Texas"*

SCENIC ROUTE: *Route 170 winds along the Rio Grande and into the stunning scenery of Big Bend National Park.*

TEXAS is the second-largest state in size and population. Huge reserves of oil lie beneath the ground, making Texas the leading state for oil production.

Texas is also the number one state in growing cotton and raising beef cattle. There are more farms in Texas than in any other state. Huge herds of cattle are raised in feedlots and on the Texas plains. The largest ranch of all, the King Ranch, is about the size of Rhode Island. Texas has as many cattle as it does people.

The name *Texas* was taken from the Indian word *Tejas,* meaning "friends." The single star on the state flag is the source of Texas' nickname, the "Lone Star State." Through the years, Texans have lived under the governments of Spain, France, Mexico, the Republic of Texas, the Confederacy, and the United States. Six Flags Over Texas, a popular amusement park in Arlington, refers to these six different governments.

"Remember the Alamo!" was a common cry in the 1830s as Texas fought for independence from Mexico. The Alamo was the chapel of an old Spanish church in San Antonio. Fewer than 200 members of the Texas army fought Mexican General Antonio López de Santa Anna's army of 5,000 soldiers between February 23 and March 6, 1836, at the Alamo. Davy Crockett and Jim Bowie were among the brave Americans who died in this bloody victory for Mexico. A month later, under the leadership of General Sam Houston, the Texans routed the Mexican Army at San Jacinto and gained their independence.

The Lyndon B. Johnson Space Center in Houston, the state's largest city, is the headquarters for American manned space flights. Astronauts are trained there. Each manned space craft that blasts off at Cape Canaveral in Florida is guided and monitored by Houston's Mission Control experts.

The capital of Texas is Austin. It was named after Stephen F. Austin, who established the first official American colony in Texas in 1822. Dallas, an important commercial and financial center, is found in the northern part of the state.

For more information about Texas call (800) 8888-TEX (888-8839)

Clockwise from lower left, the rugged Texas panhandle; the tropical Rio Grande Valley; horses and wranglers in west Texas; an Hispanic dancer in south Texas; *top,* the Texas gulf coast; *opposite page,* Dallas.

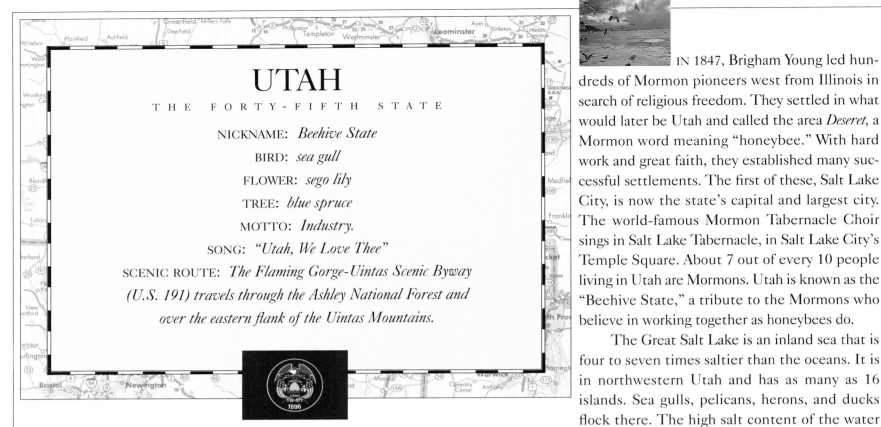

UTAH

THE FORTY-FIFTH STATE

NICKNAME: *Beehive State*

BIRD: *sea gull*

FLOWER: *sego lily*

TREE: *blue spruce*

MOTTO: *Industry.*

SONG: *"Utah, We Love Thee"*

SCENIC ROUTE: *The Flaming Gorge-Uintas Scenic Byway (U.S. 191) travels through the Ashley National Forest and over the eastern flank of the Uintas Mountains.*

IN 1847, Brigham Young led hundreds of Mormon pioneers west from Illinois in search of religious freedom. They settled in what would later be Utah and called the area *Deseret*, a Mormon word meaning "honeybee." With hard work and great faith, they established many successful settlements. The first of these, Salt Lake City, is now the state's capital and largest city. The world-famous Mormon Tabernacle Choir sings in Salt Lake Tabernacle, in Salt Lake City's Temple Square. About 7 out of every 10 people living in Utah are Mormons. Utah is known as the "Beehive State," a tribute to the Mormons who believe in working together as honeybees do.

The Great Salt Lake is an inland sea that is four to seven times saltier than the oceans. It is in northwestern Utah and has as many as 16 islands. Sea gulls, pelicans, herons, and ducks flock there. The high salt content of the water keeps human swimmers afloat.

Utah has other large deposits of salt—enough to supply the entire world for a thousand years. Near Wendover in western Utah are the Bonneville Salt Flats, rock-hard salt beds which are used as a speedway for racing cars. World land-speed records are set and broken in this flat desert area.

Only Arizona outranks Utah in the production of copper. In Bingham Canyon, not far from Salt Lake City, is one of the world's biggest copper mines. Other important minerals produced in the state are gold, silver, and uranium.

West Valley City, Provo, and Sandy City are Utah's largest cities. All are located south of Salt Lake City.

Bryce Canyon National Park, in southern Utah, is known for its spectacular rock formations. Over millions of years, desert winds and rains have shaped the canyon rocks into colorful creations. Utah's other national parks—Zion, Canyonlands, and Arches—also offer beautiful settings. Monument Valley, in the southeastern part of the state, has red sandstone formations as tall as 1,000 feet (300 meters). At dusk, one formation known as the "totem pole" casts a shadow 35 miles (56 kilometers) long!

Utah was named for the Ute Indian tribe, one of several that lived in the region. At Hovenweep National Monument, Native American homes dating back more than 600 years can be seen.

For more information about Utah call (800) 200-1160

Clockwise from lower left, Rainbow Bridge National Monument; Tower Arch, Klondike Bluff, Arches National Park; Bryce Canyon National Park; Salt Lake Temple; Zion National Park; Sundance Ski Resort; *top,* Great Salt Lake.

VERMONT is the leading state in the East for skiing. Running down the center of Vermont are its famous Green Mountains. Many of the state's best-known ski resorts, such as Killington and Camels Hump, are in the Green Mountains. The tallest peak in Vermont, Mount Mansfield, is also there. Located in the northern part of the state, it is 4,393 feet (1,339 meters) high.

Vermont's name comes from two French words, *vert mont*, meaning "green mountain." Not surprisingly, Vermont's nickname is the "Green Mountain State."

A Vermont military force known as the Green Mountain Boys was formed in 1770. During the Revolutionary War, Ethan Allen led the Green Mountain Boys and surprised sleeping British soldiers at Fort Ticonderoga in May 1775. The fort was captured without a shot being fired.

The largest stone-finishing plant in the world is in Barre. Huge granite blocks are cut, polished, and carved there. Granite, Vermont's leading mineral product, is taken from the large quarries near Barre. Vermont also has large deposits of marble, which were used to build the Lincoln Memorial and Supreme Court Building in Washington, D.C.

About 75 percent of Vermont is forests, and the state tree is the sugar maple. Early Native Americans first showed Vermont settlers how to tap this tree for the sweet sap inside. Today, Vermont is the leading producer of maple syrup in the United States; more than half a million gallons (over 1.9 million liters) are made each year in the state. From late February through April, workers boil off 35 gallons (132 liters) of sugar maple sap to make each gallon of maple syrup.

In north-central Vermont is the capital, Montpelier. The largest city in Vermont, Burlington, is in the northwestern part of the state. It is the only city in Vermont with more than 30,000 people. Vermont has fewer people than all the states except Wyoming.

For more information about Vermont call (800) VERMONT (837-6668)

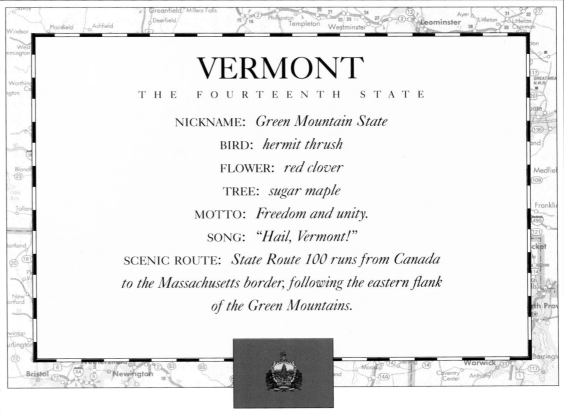

VERMONT
THE FOURTEENTH STATE

NICKNAME: *Green Mountain State*

BIRD: *hermit thrush*

FLOWER: *red clover*

TREE: *sugar maple*

MOTTO: *Freedom and unity.*

SONG: *"Hail, Vermont!"*

SCENIC ROUTE: *State Route 100 runs from Canada to the Massachusetts border, following the eastern flank of the Green Mountains.*

Top, East Corinth; *right*, Green River; *far right*, sugaring in Cabot; *top right*, Vermont's beautiful fall foliage.

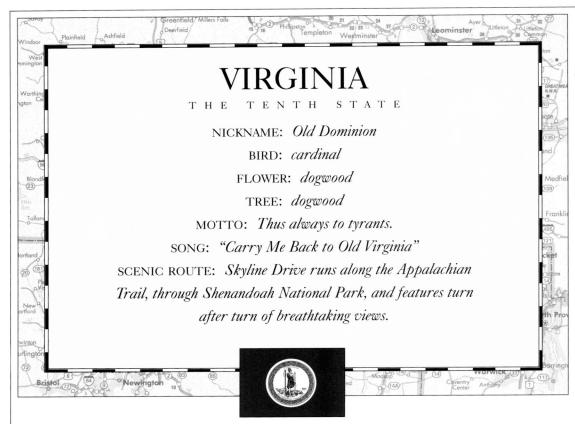

VIRGINIA

T H E   T E N T H   S T A T E

NICKNAME: *Old Dominion*

BIRD: *cardinal*

FLOWER: *dogwood*

TREE: *dogwood*

MOTTO: *Thus always to tyrants.*

SONG: *"Carry Me Back to Old Virginia"*

SCENIC ROUTE: *Skyline Drive runs along the Appalachian Trail, through Shenandoah National Park, and features turn after turn of breathtaking views.*

VIRGINIA was named by Sir Walter Raleigh in honor of England's Queen Elizabeth I, the "Virgin Queen." King Charles II nicknamed Virginia the "Old Dominion" because of its loyalty to the British royal family during the English Civil War in the middle of the 17th century.

Virginia has played an important role in U.S. history. America's first permanent English settlement was founded at Jamestown, Virginia, in 1607. In 1775, Patrick Henry spoke his famous words, "Give me liberty or give me death!" at St. John's Church in Richmond, as part of a speech he gave before the Second Virginia Convention held there. In 1781, the Revolutionary War ended when Lord Cornwallis and his British troops surrendered to George Washington and his American colonial army at Yorktown.

In the mid 1800s, many of the greatest battles of the Civil War were fought on Virginia soil. Famous confederate victories took place at Manassas, Fredericksburg, and Chancellorsville. In 1865, the Civil War ended when General Robert E. Lee surrendered to General Ulysses S. Grant at the courthouse in Appomattox.

Eight presidents were born in Virginia, earning it another nickname, "Mother of Presidents," a nickname it shares with Ohio. Those presidents were George Washington, Thomas Jefferson, James Madison, James Monroe, William Henry Harrison, John Tyler, Zachary Taylor, and Woodrow Wilson. Mount Vernon, Washington's home, and Monticello, Jefferson's home, are famous historical sites.

Northern Virginia, on the outskirts of Washington, D.C., is home to thousands of federal government workers. The Pentagon in Arlington is the largest office building in the world, employing more than 23,000 people.

Richmond is the capital of Virginia. It was also the capital of the Confederacy from 1861 to 1865. Norfolk and Virginia Beach, a popular resort, are Virginia's largest cities.

Tobacco is the most valuable crop grown in Virginia. Coal is its leading mining product. The state is also among the leaders for crab and oyster production.

For more information about Virginia call (800) 847-4882

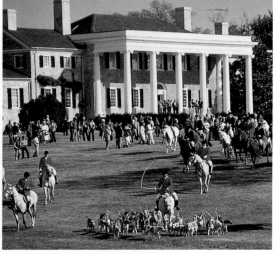

Clockwise from lower left, Mount Vernon, near Alexandria; Great Falls Park on the Potomac River; Virginia Beach; the Blue Ridge Hunt, Carter Hall; Williamsburg; *top,* Marby Hill on the Blue Ridge Parkway.

THE state of Washington was named in honor of George Washington. It is split, north to south, by the scenic Cascade Mountains, which were formed only five million years ago—young, as mountains go. Since then, glaciers have continued to carve away sharp ridges. Mount Rainier, at 14,410 feet (4,392 meters), is the highest peak in the Cascade Mountains and an inactive volcano. Another volcano in the Cascade Mountains, Mount St. Helens, erupted on May 18, 1980, causing 57 deaths and billions of dollars in damage. About 172 square miles (445 square kilometers) of the devastated area are included in Mount St. Helens National Volcanic Monument, established in 1982.

Lands west of the Cascade Mountains have a mild, moist climate. Thick forests of fir and other evergreen trees blanket the area. Lumbering is big business in the "Evergreen State." The forests of Olympic Peninsula are among the rainiest places in the world. An average of about 135 inches (343 centimeters) of rain, snow, and other natural moisture falls on some parts of this peninsula each year.

Puget Sound is a big bay that is almost completely enclosed by land. The beautiful San Juan Islands lie at the mouth of this bay. Most of Washington's largest cities are also located on Puget Sound. The largest city of all, Seattle, is an important port and a center for fishing. Washington is famous for its delicious salmon. From the observation deck at the top of Seattle's 607-foot (185-meter) Space Needle, visitors have a wonderful view of the city and the nearby mountains.

East of the Cascades, Washington lands are very dry. Dozens of dams have been built to harness the Columbia River and irrigate new farmlands. The Grand Coulee Dam in northeastern Washington is the largest concrete dam in the United States. The state's irrigated lands are known for their many orchards, which have helped make Washington the leading U.S. producer of apples.

Spokane and Tacoma are the largest cities after Seattle. Olympia, the capital, is in western Washington.

For more information about Washington call (800) 544–1800

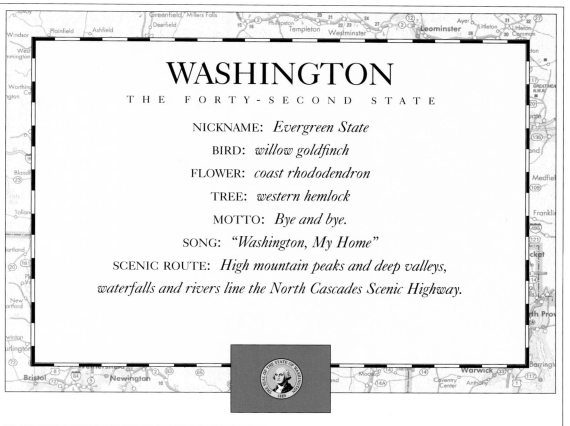

WASHINGTON
THE FORTY-SECOND STATE

NICKNAME: *Evergreen State*

BIRD: *willow goldfinch*

FLOWER: *coast rhododendron*

TREE: *western hemlock*

MOTTO: *Bye and bye.*

SONG: *"Washington, My Home"*

SCENIC ROUTE: *High mountain peaks and deep valleys, waterfalls and rivers line the North Cascades Scenic Highway.*

Top, the Space Needle, Seattle; *right,* Puget Sound; *far right,* Meta Lake, Mount St. Helens National Volcanic Monument; *top right,* Paradise Valley, Mount Rainier National Park.

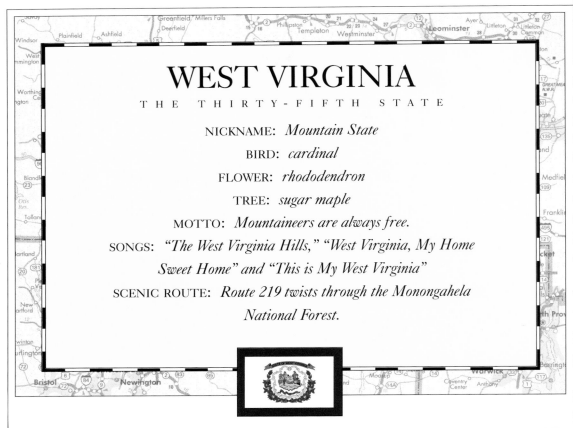

WEST VIRGINIA

THE THIRTY-FIFTH STATE

NICKNAME: *Mountain State*

BIRD: *cardinal*

FLOWER: *rhododendron*

TREE: *sugar maple*

MOTTO: *Mountaineers are always free.*

SONGS: *"The West Virginia Hills," "West Virginia, My Home Sweet Home" and "This is My West Virginia"*

SCENIC ROUTE: *Route 219 twists through the Monongahela National Forest.*

WEST VIRGINIA was originally part of Virginia but broke away to become a separate state during the Civil War. Almost 80 percent of West Virginia is forested. It is so hilly that the only way workers building an airport at Charleston could make flat runways for planes to land was to cut off mountaintops and pour extra dirt into the valleys. Except for occasional river valleys, West Virginia has almost no level land at all. The rough, uneven surface led to its nickname of "Mountain State."

Only Kentucky and Wyoming outrank West Virginia in the production of coal in the United States. Deposits of soft coal, called bituminous coal, can be found under nearly 50 percent of the state.

West Virginia also has rich deposits of silica and natural gas. Both are used in glassmaking, one of its most important industries. Most of the glass marbles made in America come from the area around Parkersburg.

Harpers Ferry National Historical Park sits where the Shenandoah and Potomac rivers meet. It lies in the northeastern part of West Virginia, close to the Maryland border. In 1859, abolitionist John Brown and his followers captured the U.S. arsenal there. He hoped to start a slave revolt, but it did not happen. Brown himself was captured, found guilty of murder and revolt against his country, and hanged.

Charleston is West Virginia's capital and largest city. It rests along the banks of the Kanawha River, which flows into the Ohio River. The state's next three largest cities are Huntington, Wheeling, and Parkersburg.

Outdoor enthusiasts enjoy hunting, fishing, hiking, and camping in the George Washington and Monongahela national parks. Throughout the state, various levels of white-water rafting are offered on seven different rivers. Blackwater Falls State Park, near Davis, is known for its beautiful scenery.

For more information about West Virginia call (800) CALL-WVA (225-5982)

Clockwise from lower left, Harper's Ferry; Endless Wall, New River Gorge; Seneca Rocks; Blackwater Falls; Pillow Rock Rapid, Gauley River; *top,* Beckley Coal Mine.

A POPULAR nickname for Wisconsin is "America's Dairyland." With about two million milk cows housed on its dairy farms, Wisconsin leads all states in the production of milk. It also leads the country in the production of butter and is among the top producers of cheese.

The name *Wisconsin* is a Native American word believed to mean "gathering of the waters." The state has more than 10,000 streams, 15,000 lakes, and numerous rivers. Lake Winnebago, in the eastern part of the state, is Wisconsin's largest lake, filling 215 square miles (557 square kilometers). The state is bordered by the St. Croix and Mississippi rivers, Green Bay, and lakes Superior and Michigan. The Great Lakes provide natural air conditioning as well as additional opportunities for fishing, boating, and swimming.

Northern Wisconsin is heavily forested, which enables lumber and paper production to be big businesses there. The state is among the country's leaders for paper production. Also in the northern part of the state is Timm's Hill, Wisconsin's highest point at 1,952 feet (595 meters).

In the 1820s, people flocked to southwestern Wisconsin to mine the lead deposits. Many miners did not take the time to build houses. Instead, they lived in caves or shelters dug in the hillsides. This reminded some people of what badgers do—dig into the earth for places to live—and the miners were soon nicknamed "badgers." Since that time, Wisconsin has been known as the "Badger State."

Many of Wisconsin's early settlers came from Scandinavia. Not far from Madison, the state's capital and second-largest city, is Little Norway. There, visitors will find a Norwegian village built as it would have been in the early 1800s. Nearby New Glarus is known as "Swissconsin" because of its many Swiss buildings and celebrations.

Milwaukee is Wisconsin's largest city and is famous for its beer production. The third-largest city, Green Bay, is the home of the National Football League's Green Bay Packers.

For more information about Wisconsin call (800) 432-8747

WISCONSIN
THE THIRTIETH STATE

NICKNAME: *Badger State*

BIRD: *robin*

FLOWER: *wood violet*

TREE: *sugar maple*

MOTTO: *Forward.*

SONG: *"On, Wisconsin!"*

SCENIC ROUTE: *Route 55 is a beautiful drive along Lake Winnebago, the largest lake in the state.*

WISCONSIN
1848

Top, Wisconsin's famous dairy cows; *clockwise from lower left,* cross-country skiing; the capitol in Madison; a scenic Wisconsin view; Hawk's Beak, Wisconsin Dell.

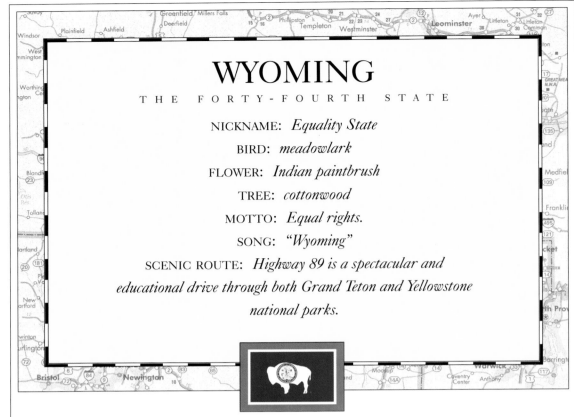

WYOMING

THE FORTY-FOURTH STATE

NICKNAME: *Equality State*

BIRD: *meadowlark*

FLOWER: *Indian paintbrush*

TREE: *cottonwood*

MOTTO: *Equal rights.*

SONG: *"Wyoming"*

SCENIC ROUTE: *Highway 89 is a spectacular and educational drive through both Grand Teton and Yellowstone national parks.*

WYOMING is a sprawling state of beautiful parks, large ranches, and abundant natural resources. Yellowstone National Park, tucked in the northwest corner of the state, is the oldest national park in the world and the largest wildlife preserve in the United States, covering 2,219,785 acres (898,315 hectares). Established in 1872, the park got its name from the yellow rocks lining the Yellowstone River section north of it. Yellowstone has thousands of hot springs and over 200 geysers. Old Faithful is the most famous geyser in the park, erupting to a height of more than 100 feet (30 meters) about every 73 minutes.

Grand Teton National Park is known for its beautiful lakes and spectacular mountain scenery. In this protected park area, many different types of wildlife can be seen. Another interesting place to visit is Devils Tower National Monument in northeastern Wyoming. This unusual volcanic tower juts 1,280 feet (390 meters) into the air. In 1906, it was designated as the nation's first national monument.

Wyoming is known as the "Equality State" because it was the first state to give women the right to vote. That right was granted in 1869 while Wyoming was still a territory. Today, Wyoming is the ninth largest state in area, but ranks last in population.

Wyoming is a Delaware Indian word meaning "on the great plain." Today, sheep and cattle ranches stretch across Wyoming's plains where Sioux and Cheyenne Indians once hunted, and outlaw Butch Cassidy later traveled with his Hole-in-the-Wall gang. Raising beef cattle is very important to the state's economy. Wyoming is behind only Texas and California in the production of sheep and wool in the U.S. Other products Wyoming is known for include coal, oil, and uranium.

Reminders of Wyoming's rich heritage are everywhere. Visitors to the Buffalo Bill Historical Center in Cody can learn about the famous buffalo hunter who helped found that city. Fort Laramie in southeastern Wyoming was a fur-trading post and an important stop for wagon trains traveling on the Oregon Trail. It has been restored to look as it did in pioneer days. Each July, the city of Cheyenne, Wyoming's capital and largest city, hosts a Frontier Day celebration that includes parades, rodeos, and Indian dances.

For more information about Wyoming call (800) 225-5996

Clockwise from lower left, Shoshone Boys, Wind River Reservation; Devil's Tower National Monument; trail ride near the Tetons; Old Faithful, Yellowstone National Park; Old Trail Town, Cody; *top,* Cheyenne Frontier Days.

WASHINGTON, D.C., the capital of the United States and the headquarters for its government, lies on the east bank of the Potomac River between Virginia and Maryland. It is one of America's most beautiful and historic cities.

The District of Columbia was laid out by French engineer Pierre Charles L'Enfant on land donated by Maryland. Contests were held to select the best designs for major buildings in the city. On September 18, 1793, President George Washington laid the cornerstone for the Capitol, and a city began to form where swamplands had once been.

Today, the nation's capital is filled with statues, memorials, and buildings of historical interest. Near the center of the city is Capitol Hill, which includes the Capitol, where Congress meets and makes laws for the nation, and the Supreme Court Building, where the nation's constitution and laws are interpreted. The Library of Congress is also there. It may be the largest library in the world, with over 84 million items housed inside.

West of Capitol Hill are three monuments honoring presidents George Washington, Thomas Jefferson, and Abraham Lincoln. The Washington Monument is a pillar rising nearly 555 feet (169 meters), making it the highest structure in the city. The Jefferson Memorial is a domed building that houses a bronze statue of the country's third president. The Lincoln Memorial is a white-marble monument featuring a large statue of the Civil War president sitting in a chair.

The National Museum of American History and the National Museum of Natural History are nearby. They are part of the great Smithsonian Institution and contain displays ranging from historical automobiles to the Hope Diamond.

The White House, home and office to the President of the United States, is located at 1600 Pennsylvania Avenue in the center of the city. This gracious mansion has 132 rooms. Every president, with the exception of George Washington, has lived there.

For more information about Washington, D.C. call (202) 789-7000

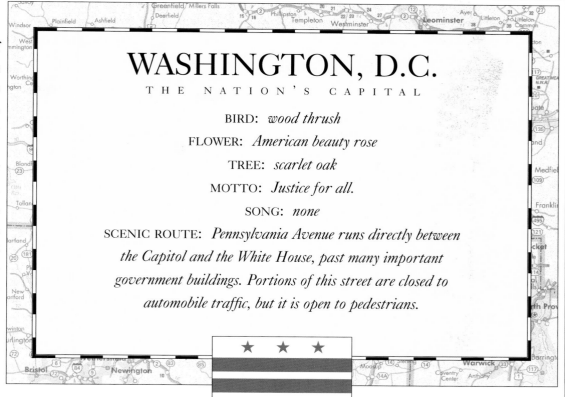

WASHINGTON, D.C.
THE NATION'S CAPITAL

BIRD: *wood thrush*

FLOWER: *American beauty rose*

TREE: *scarlet oak*

MOTTO: *Justice for all.*

SONG: *none*

SCENIC ROUTE: *Pennsylvania Avenue runs directly between the Capitol and the White House, past many important government buildings. Portions of this street are closed to automobile traffic, but it is open to pedestrians.*

★ ★ ★

Top, the White House; *right,* the Jefferson Memorial; *far right,* the Washington Monument; *top right,* the Capitol Building.

IN addition to the 50 states, the United States includes a number of widespread islands in the world. U.S. flags fly more than 10,000 miles (16,093 kilometers) apart, from Puerto Rico and the Virgin Islands in the Caribbean to the Pacific island of Guam.

The island of Puerto Rico is nearly 1,000 miles (1,600 kilometers) southeast of Florida. The people of Puerto Rico speak mostly Spanish and are U.S. citizens. Puerto Rico's beautiful beaches and historic buildings attract many tourists each year. El Morro Fortress near the Bay of San Juan was built by the Spaniards between 1539 and 1787.

The Virgin Islands are two groups of small islands owned by the United States and Great Britain. The Virgin Islands of the United States are about 40 miles (64 kilometers) east of Puerto Rico. Only St. Croix (the largest), St. John, and St. Thomas have people living on them permanently, although there are other, smaller islands. Tourism is the major industry on the Virgin Islands.

More U.S. territories are on the other side of the world. American Samoa is made up of seven Pacific islands located 4,800 miles (7,720 kilometers) southwest of San Francisco. The largest island is Tutuila. Most American Samoans are Polynesians, and many of them work at fishing and growing coconuts.

Other Pacific islands owned by the U.S. include the Marianas. The total population of these islands is about 138,400, but nearly 120,000 people live on the largest Mariana island, Guam. An important U.S. air and naval base is located there.

During World War II, Guam and several other islands played key roles in Pacific battles. In June 1942, the Battle of Midway took place in the waters surrounding Midway Island, another U.S. possession which is located about 1,300 miles (2,090 kilometers) northwest of Honolulu, Hawaii. This battle was the first major naval victory by the U.S. over Japan.

For more information about Puerto Rico call (800) 223-6530

Top, El Morro Fortress, San Juan, Puerto Rico; *right,* the east coastline, American Samoa; *top right,* Cruz Bay, U.S. Virgin Islands; *opposite page,* San Juan Jeronimo Fort, Puerto Rico.

U.S. POSSESSIONS

PUERTO RICO *was ceded to the U.S. during the Spanish-American War of 1898, and became a territory in 1917.*

The U.S. VIRGIN ISLANDS *were purchased from Denmark for $25 million in 1917.*

The Treaty of Berlin, signed in 1899, acknowledged the United States' rights over the islands now known as AMERICAN SAMOA, *but Congress did not formally accept sovereignty until 1929.*

GUAM *was acquired from Spain in 1898, and became a territory in 1950. The remaining* MARIANAS ISLANDS *became part of the United States in 1986.*

	RANKING OF THE STATES BY AREA		
	STATE	SQUARE MILES	SQUARE KILOMETERS
1	ALASKA	591,004	1,530,700
2	TEXAS	266,807	691,030
3	CALIFORNIA	158,706	411,049
4	MONTANA	147,046	380,848
5	NEW MEXICO	121,593	314,295
6	ARIZONA	114,000	295,260
7	NEVADA	110,561	286,352
8	COLORADO	104,091	269,595
9	WYOMING	97,809	253,326
10	OREGON	97,073	251,419
11	UTAH	84,899	219,887
12	MINNESOTA	84,402	218,601
13	IDAHO	83,564	216,432
14	KANSAS	82,264	213,063
15	NEBRASKA	77,355	200,350
16	SOUTH DAKOTA	77,116	199,730
17	NORTH DAKOTA	70,702	183,119
18	OKLAHOMA	69,956	181,186
19	MISSOURI	69,697	180,516
20	WASHINGTON	68,139	176,479
21	GEORGIA	58,910	152,576
22	FLORIDA	58,664	151,939
23	MICHIGAN	58,527	151,586
24	ILLINOIS	56,345	145,934
25	IOWA	56,275	145,753
26	WISCONSIN	56,154	145,438
27	ARKANSAS	53,187	137,754
28	NORTH CAROLINA	52,669	136,413
29	ALABAMA	51,705	133,915
30	NEW YORK	49,108	127,189
31	LOUISIANA	47,752	123,677
32	MISSISSIPPI	47,689	123,515
33	PENNSYLVANIA	45,308	117,348
34	TENNESSEE	42,114	109,152
35	OHIO	41,330	107,044
36	VIRGINIA	40,767	105,586
37	KENTUCKY	40,395	104,623
38	INDIANA	36,185	93,720
39	MAINE	33,265	86,156
40	SOUTH CAROLINA	31,113	80,582
41	WEST VIRGINIA	24,231	62,759
42	MARYLAND	10,460	27,092
43	VERMONT	9,614	24,900
44	NEW HAMPSHIRE	9,297	24,032
45	MASSACHUSETTS	8,284	21,456
46	NEW JERSEY	7,787	20,169
47	HAWAII	6,471	16,759
48	CONNECTICUT	5,018	12,997
49	DELAWARE	2,044	5,295
50	RHODE ISLAND	1,212	3,140

	RANKING OF THE STATES BY POPULATION*	
1	CALIFORNIA	31,431,000
2	TEXAS	18,378,000
3	NEW YORK	18,169,000
4	FLORIDA	13,953,000
5	PENNSYLVANIA	12,052,000
6	ILLINOIS	11,752,000
7	OHIO	11,102,000
8	MICHIGAN	9,496,000
9	NEW JERSEY	7,904,000
10	NORTH CAROLINA	7,070,000
11	GEORGIA	7,055,000
12	VIRGINIA	6,552,000
13	MASSACHUSETTS	6,041,000
14	INDIANA	5,752,000
15	WASHINGTON	5,343,000
16	MISSOURI	5,278,000
17	TENNESSEE	5,175,000
18	WISCONSIN	5,082,000
19	MARYLAND	5,006,000
20	MINNESOTA	4,567,000
21	LOUISIANA	4,315,000
22	ALABAMA	4,219,000
23	ARIZONA	4,075,000
24	KENTUCKY	3,827,000
25	SOUTH CAROLINA	3,664,000
26	COLORADO	3,656,000
27	CONNECTICUT	3,275,000
28	OKLAHOMA	3,258,000
29	OREGON	3,086,000
30	IOWA	2,829,000
31	MISSISSIPPI	2,669,000
32	KANSAS	2,554,000
33	ARKANSAS	2,453,000
34	UTAH	1,908,000
35	WEST VIRGINIA	1,822,000
36	NEW MEXICO	1,654,000
37	NEBRASKA	1,623,000
38	NEVADA	1,457,000
39	MAINE	1,240,000
40	HAWAII	1,179,000
41	NEW HAMPSHIRE	1,137,000
42	IDAHO	1,133,000
43	RHODE ISLAND	997,000
44	MONTANA	856,000
45	SOUTH DAKOTA	721,000
46	DELAWARE	706,000
47	NORTH DAKOTA	638,000
48	ALASKA	606,000
49	VERMONT	580,000
50	WYOMING	476,000

*BASED ON 1994 CENSUS ESTIMATES.

Top: dogsledding across Alaska, the largest state.
Bottom: the famous Cliff Walk along the coast of Rhode Island, the smallest state.

Top: swimmers at Venice Beach in California, the most populated state.
Bottom: wheel ruts along the Oregon Trail in Wyoming, the least populated state.

State	Date		State	Date
DELAWARE	DECEMBER 7, 1787		MICHIGAN	JANUARY 26, 1837
PENNSYLVANIA	DECEMBER 12, 1787		FLORIDA	MARCH 3, 1845
NEW JERSEY	DECEMBER 18, 1787		TEXAS	DECEMBER 29, 1845
GEORGIA	JANUARY 2, 1788		IOWA	DECEMBER 28, 1846
CONNECTICUT	JANUARY 9, 1788		WISCONSIN	MAY 29, 1848
MASSACHUSETTS	FEBRUARY 6, 1788		CALIFORNIA	SEPTEMBER 9, 1850
MARYLAND	APRIL 28, 1788		MINNESOTA	MAY 11, 1858
SOUTH CAROLINA	MAY 23, 1788		OREGON	FEBRUARY 14, 1859
NEW HAMPSHIRE	JUNE 21, 1788		KANSAS	JANUARY 29, 1861
VIRGINIA	JUNE 25, 1788		WEST VIRGINIA	JUNE 20, 1863
NEW YORK	JULY 26, 1788		NEVADA	OCTOBER 31, 1864
NORTH CAROLINA	NOVEMBER 21, 1789		NEBRASKA	MARCH 1, 1867
RHODE ISLAND	MAY 29, 1790		COLORADO	AUGUST 1, 1876
VERMONT	MARCH 4, 1791		NORTH DAKOTA	NOVEMBER 2, 1889
KENTUCKY	JUNE 1, 1792		SOUTH DAKOTA	NOVEMBER 2, 1889
TENNESSEE	JUNE 1, 1796		MONTANA	NOVEMBER 8, 1889
OHIO	MARCH 1, 1803		WASHINGTON	NOVEMBER 11, 1889
LOUISIANA	APRIL 30, 1812		IDAHO	JULY 3, 1890
INDIANA	DECEMBER 11, 1816		WYOMING	JULY 10, 1890
MISSISSIPPI	DECEMBER 10, 1817		UTAH	JANUARY 4, 1896
ILLINOIS	DECEMBER 3, 1818		OKLAHOMA	NOVEMBER 16, 1907
ALABAMA	DECEMBER 14, 1819		NEW MEXICO	JANUARY 6, 1912
MAINE	MARCH 15, 1820		ARIZONA	FEBRUARY 14, 1912
MISSOURI	AUGUST 10, 1821		ALASKA	JANUARY 3, 1959
ARKANSAS	JUNE 15, 1836		HAWAII	AUGUST 21, 1959

❧ Seven different states have the cardinal as their state bird: Illinois, Indiana, Kentucky, North Carolina, Ohio, Virginia, and West Virginia.

❧ The most common state tree is the pine tree. It is the state tree of Alabama, Arkansas, Idaho, Maine, Michigan, Minnesota, Montana, New Mexico, and North Carolina.

❧ The rose is the most common state flower. Four states have it: Georgia, Iowa, New York, and North Dakota. Washington, D.C. also has it.

❧ The northern boundary of Delaware is shaped like a semicircle. It is the only state with a round border.

❧ Four Corners is the only point in the United States where four states meet. Arizona, Colorado, New Mexico, and Utah all come together there.

❧ Colorado has more than 1,000 mountains over 10,000 feet (3,000 meters) high.

❧ The President of the United States lives in the White House in Washington, D.C., yet that city's residents could not vote for a U.S. president until the 1964 election.

❧ Pennsylvania had the first bank, first lending library, first zoo, first hospital, and first paved highway in the United States.

❧ The lake with the longest known name is Lake Chargoggagoggmanchaugagogg-chabunagungamaugg, in Massachusetts. The Indian name means "You fish on your side. I fish on my side. Nobody fishes in the middle." It is also called Lake Webster.

❧ The world's largest manmade hole is Kennecott Copper's mine in Bingham Canyon, Utah.

❧ Alaska is closer to the Soviet Union than it is to any of the other 49 states. Alaska's Little Diomede Island is only 2 miles (approximately 4 kilometers) away from the Russian island of Big Diomede. The mainlands are 51 miles (82 kilometers) apart, yet Alaska is 500 miles (800 kilometers) from Washington, the nearest state.

❧ London Bridge, of nursery rhyme fame, now stands at Lake Havasu City in Arizona. It was taken apart in England, moved to the United States, and rebuilt there in 1971.

❧ New York City was the first capital of the United States. It was capital from 1775 until 1790. Philadelphia was the nation's capital from 1790 until 1800.

❧ The world's longest bridge is the Lake Ponchartrain Causeway in New Orleans, Louisiana. Its total length is about 29 miles (47 kilometers).

❧ The first woman elected to the U.S. Congress was Jeannette Rankin. The people of Montana voted her into office in 1917.

❧ The largest living object in the world is a sequoia tree in northern California. It weighs more than 2,000 short tons (1,814 metric tons).

❧ Natural Bridge, a large stone formation created by water in western Virginia, was once owned by Thomas Jefferson and was surveyed by George Washington. Washington's carved initials can still be seen on it today.

❧ Alaska's coastline is longer than the coastlines of all the other states put together.

❧ As many as 4,000 bathtubs' worth of water pours over New York's Niagara Falls every second.

❧ The Gateway Arch in St. Louis, with a height of 630 feet (192 meters) is the tallest manmade monument in the United States.

❧ The world's largest chocolate factory is in Hershey, Pennsylvania.

❧ The highest tides off the U.S. mainland are at Passamaquoddy Bay, Maine. There, the waters rise 26 feet (8 meters).

❧ The world's first atomic bomb was detonated in the deserts of New Mexico in 1945. The blast of heat was so intense that desert sands turned into glass.

❧ Michigan is the only state separated into two parts.

❧ The five Great Lakes contain a fourth of all the fresh water in the world. This is enough water to cover the United States to a depth of 12 feet (4 meters).

❧ The Little League World Series is held each year in Williamsport, Pennsylvania.

❧ Texas is 220 times as big as Rhode Island.

❧ Two of every five towns in Nevada are ghost towns. Most of these are abandoned mining camps.

❧ The largest manmade forests in the world are found in Nebraska. More than 22,000 acres (8,800 hectares) of trees were planted in the dry Sand Hills region to help prevent wind erosion.

❧ Basketball was first played in Springfield, Massachusetts, in 1891. Peach baskets were nailed to a wall, and the ball had to be taken out after every score.

❧ NORTHERNMOST POINT: Point Barrow, Alaska

❧ SOUTHERNMOST POINT: Ka Lae (South Cape) on the island of Hawaii

❧ EASTERNMOST POINT: West Quoddy Head, Maine

❧ WESTERNMOST POINT: Cape Wrangell, Alaska

❧ HIGHEST POINT: Mount McKinley in Alaska; 20,320 feet (6,194 meters)

❧ LOWEST POINT: Death Valley, California; 282 feet (86 meters) below sea level

❧ LARGEST LAKE: Lake Superior; 31,700 square miles (82,420 square kilometers)

❧ DEEPEST LAKE: Crater Lake in Oregon; 1,932 feet (589 meters)

❧ LONGEST RIVER: Mississippi River; 2,348 miles (3,779 kilometers)

❧ LARGEST CANYON: Grand Canyon in Arizona; 277 miles (446 kilometers) long, 4-18 miles (6.4-29 kilometers) wide, and 1 mile (1.6 kilometers) deep

❧ DEEPEST CANYON: Hells Canyon in Idaho; 7,900 feet (2,408 meters)

❧ LARGEST LIVING THING: General Sherman sequoia; 275 feet (84 meters) tall and 103 feet (31.4 meters) around

❧ TALLEST BUILDING: Sears Tower in Chicago, Illinois; 1,454 feet (443 meters)

❧ OLDEST NATIONAL PARK: Yellowstone National Park; established in 1872 in Wyoming, Montana, and Idaho

❧ LARGEST NATIONAL PARK: Wrangell-St. Elias National Park in Alaska; 13,018 square miles (33,847 square kilometers)

❧ HIGHEST TEMPERATURE: 134 degrees Fahrenheit (56.7 degrees Celsius) at Death Valley, California, on July 10, 1913

❧ LOWEST TEMPERATURE: Minus 80 degrees Fahrenheit (minus 62.2 degrees Celsius) at Prospect Creek, Alaska, on January 23, 1971

❧ WETTEST SPOT: Mt. Waialeale, Hawaii; average annual rainfall of 460 inches (11,684 millimeters)

❧ DRIEST SPOT: Death Valley, California; average rainfall of 1.35 inches (34.3 millimeters)

❧ HEAVIEST SNOWFALL: 1,224.5 inches (3,110 centimeters) at Paradise Ranger Station on Mt. Ranier in Washington during the winter of 1971–72

❧ STRONGEST WIND: 231 miles per hour (372 kilometers per hour) on Mt. Washington in New Hampshire in 1934

MAJOR PRODUCTS AND INDUSTRIES OF THE UNITED STATES

STATE	MANUFACTURING	FARMING AND FISHING	MINING
ALABAMA	Primary metals, paper products, chemicals, textiles, clothing, food products	Poultry, beef cattle, soybeans, eggs, peanuts, hogs, cotton, shrimp, oysters	Coal, petroleum, natural gas, stone, iron ore, limestone
ALASKA	Food products, paper products, lumber and wood products, petroleum and coal products	Greenhouse and nursery products, milk, potatoes, hay, salmon, crabs, halibut, shrimp	Petroleum, natural gas, sand and gravel, stone, gold
ARIZONA	Nonelectric machinery, electric machinery, transportation equipment, primary metals	Beef cattle, cotton, milk, hay	Copper, molybdenum, coal, sand and gravel
ARKANSAS	Food products, electric machinery, lumber and wood products, paper products, chemicals	Soybeans, poultry, rice, beef cattle, eggs, cotton	Petroleum, bromine, natural gas, stone, sand and gravel, bauxite, diamonds
CALIFORNIA	Transportation equipment, electric machinery, food products, nonelectric machinery, metal products, printed materials	Beef cattle, milk, grapes, cotton, greenhouse and nursery products, almonds, tomatoes, hay, lettuce, tuna, salmon, crabs	Petroleum, natural gas, sand and gravel, boron
COLORADO	Food products, instruments, nonelectric machinery, metal products, printed materials, transportation equipment	Beef cattle, wheat, corn, milk, hay, sheep, beans	Petroleum, natural gas, coal, molybdenum and other metals
CONNECTICUT	Transportation equipment, nonelectric machinery, metal products, electric machinery, chemicals, instruments, submarines	Milk, eggs, greenhouse and nursery products, tobacco, beef cattle, apples, lobsters, oysters, clams	Stone, sand and gravel
DELAWARE	Chemicals, food products, metal products, paper products, printed materials, rubber and plastic products	Broilers, soybeans, corn, milk, crabs, clams	Sand and gravel, magnesium compounds
FLORIDA	Food products, electric machinery, transportation equipment, chemicals, printed materials	Grapefruit, oranges, beef cattle, greenhouse and nursery products, milk, tomatoes, shrimp, lobsters	Phosphate rock, petroleum, stone
GEORGIA	Textiles, food products, transportation equipment, chemicals, clothing, paper products	Broilers, peanuts, eggs, soybeans, beef cattle, hogs, cotton, pecans, peaches, crabs, shrimp	Clays, stone, sand and gravel
HAWAII	Food products, printed materials, clothing, transportation equipment, chemicals	Sugar cane, pineapples, beef cattle, flowers, coffee, macadamia nuts, tuna, snapper	Stone, sand and gravel
IDAHO	Lumber and wood products, food products, nonelectric machinery, chemicals	Beef cattle, wheat, potatoes, milk, barley, sugar beets	Silver, phosphate rock, lead, zinc, gemstones, gold
ILLINOIS	Nonelectric machinery, food products, electric machinery, metal products, chemicals	Corn, soybeans, oats, wheat, hogs, beef cattle, milk	Coal, petroleum, stone

STATE		MANUFACTURING	FARMING AND FISHING	MINING
INDIANA		Metals, transportation equipment, electric machinery, nonelectric machinery, chemicals	Corn, wheat, soybeans, hogs, beef cattle, milk	Coal, stone
IOWA		Nonelectric machinery, food products, electrical machinery, chemicals	Hogs, beef cattle, corn, soybeans, milk, eggs	Stone, sand and gravel
KANSAS		Transportation equipment (especially airplanes), nonelectric machinery, food products, chemicals	Beef cattle, wheat, sorghum grain, hogs, corn, soybeans, milk	Petroleum, natural gas
KENTUCKY		Nonelectric machinery, transportation equipment, electric machinery, food products, tobacco products	Tobacco, beef cattle, soybeans, milk, corn, hogs, horse breeding	Coal, stone, petroleum, natural gas, sand and gravel
LOUISIANA		Chemicals, petroleum and coal products, food products, paper products, transportation equipment	Soybeans, beef cattle, rice, cotton, milk, sugar cane, shrimp, menhaden	Natural gas, petroleum, sulfur, salt, sand and gravel, stone
MAINE		Paper products, lumber and wood products, leather products, food products	Potatoes, eggs, milk, lobsters, clams	Sand and gravel, garnet
MARYLAND		Food products, metals, electric equipment, chemicals	Broilers, milk, corn, soybeans, beef cattle, tobacco, crabs, clams, oysters	Coal, stone, sand and gravel, clays
MASSACHUSETTS		Nonelectric machinery, electric equipment, instruments, printed materials, transportation equipment, food products	Milk, greenhouse and nursery products, cranberries, eggs, apples, beef cattle, cod, scallops, flounder, lobsters, haddock	Stone, sand and gravel, clays
MICHIGAN		Transportation equipment (especially cars), electric machinery, chemicals, food products	Milk, beef cattle, corn, soybeans, hogs, whitefish, chubs	Copper, iron ore, petroleum, natural gas, sand and gravel
MINNESOTA		Nonelectric machinery, food products, metal products, electric machinery, printed materials, paper products	Milk, soybeans, beef cattle, corn, hogs, wheat, catfish, walleye, carp	Iron ore, sand and gravel, stone
MISSISSIPPI		Transportation equipment, electric machinery, lumber and wood products, food products, clothing	Soybeans, cotton, beef cattle, poultry, milk, menhaden, shrimp, red snapper	Petroleum, natural gas, sand and gravel, clays
MISSOURI		Transportation equipment, food products, chemicals, electric equipment	Beef cattle, soybeans, hogs, milk	Lead, stone
MONTANA		Lumber and wood products, food products, petroleum and coal products, printed materials	Beef cattle, wheat, barley, milk, hay, hogs, Christmas trees	Petroleum, coal, copper, natural gas, silver, gemstones
NEBRASKA		Food products, nonelectric machinery, metal products, electric machinery	Beef cattle, corn, hogs, soybeans, wheat	Petroleum, sand and gravel
NEVADA		Food products, printed materials, chemicals, lumber and wood products	Beef cattle, milk, hay, potatoes, alfalfa seed, sheep	Gold, mercury, magnesite, barite, silver, sand and gravel

MAJOR PRODUCTS AND INDUSTRIES OF THE UNITED STATES

STATE	MANUFACTURING	FARMING AND FISHING	MINING
NEW HAMPSHIRE	Nonelectric machinery, electric machinery, paper products, instruments, rubber and plastic products	Milk, eggs, greenhouse and nursery products, apples, beef cattle, lobsters, cod	Sand and gravel, stone
NEW JERSEY	Chemicals, food products, electric machinery, nonelectric machinery	Milk, greenhouse and nursery products, tomatoes, clams, menhaden, flounder	Stone, sand and gravel, zinc
NEW MEXICO	Food products, electric machinery, printed materials, lumber and wood products	Beef cattle, milk, cotton, hay	Natural gas, petroleum, uranium, copper, silver, gold, semiprecious stones
NEW YORK	Printed materials, instruments, nonelectric machinery, electric machinery, clothing, chemicals	Milk, beef cattle, apples, greenhouse and nursery products, clams, oysters	Stone, salt, sand and gravel
NORTH CAROLINA	Textiles, tobacco products, chemicals, furniture, electric machinery, food products, clothing, lumber and wood products	Tobacco, turkeys, broilers, hogs, soybeans, shrimp, menhaden, flounder	Stone, phosphate rock, sand and gravel
NORTH DAKOTA	Food products, nonelectric machinery, printed materials	Wheat, rye, beef cattle, sunflower seeds, barley, milk, sugar beets	Petroleum, coal, natural gas, sand and gravel
OHIO	Transportation equipment, nonelectric machinery, metal products, electric machinery, chemicals, rubber products	Soybeans, corn, milk, beef cattle, hogs, apples, yellow perch, white bass	Coal, petroleum, natural gas, stone, lime
OKLAHOMA	Nonelectric machinery (especially oil-field machinery), metal products, electric machinery, food products, rubber and plastic products	Beef cattle, wheat, cotton, milk, peanuts	Petroleum, natural gas, coal, stone
OREGON	Lumber and wood products, food products, paper products, nonelectric machinery, transportation equipment	Beef cattle, wheat, milk, greenhouse and nursery products, potatoes, salmon, tuna, shrimp	Sand and gravel, stone, nickel, pumice
PENNSYLVANIA	Iron and steel, nonelectric machinery, food products (especially pretzels and chocolate), electric machinery	Milk, beef cattle, mushrooms	Coal, stone
RHODE ISLAND	Jewelry and silverware, metal products, nonelectric machinery, textiles	Greenhouse and nursery products, milk, flounder, lobsters	Sand and gravel, stone
SOUTH CAROLINA	Textiles, chemicals, nonelectric machinery, clothing, paper products, electric machinery	Soybeans, tobacco, beef cattle, eggs, milk, corn, hogs, peaches, cotton, shrimp, crabs	Stone, clays
SOUTH DAKOTA	Food products, nonelectric machinery	Beef cattle, hogs, milk, wheat, corn, soybeans, oats	Gold, stone
TENNESSEE	Chemicals, food products, nonelectric machinery, electric machinery, clothing	Soybeans, beef cattle, milk, hogs, tobacco, cotton	Coal, stone, zinc

MAJOR PRODUCTS AND INDUSTRIES OF THE UNITED STATES

STATE		MANUFACTURING	FARMING AND FISHING	MINING
TEXAS		Chemicals, nonelectric machinery, petroleum and coal products, food products, transportation equipment	Beef cattle, cotton, shrimp	Petroleum, natural gas
UTAH		Metals, nonelectric machinery, transportation equipment, food products, electric machinery	Beef cattle, milk, hay, turkeys, sheep	Petroleum, coal, copper, uranium, molybdenum, gold
VERMONT		Electric machinery, nonelectric machinery, printed materials, metal products, paper products	Milk, beef cattle, eggs, apples, maple products	Stone (especially granite and marble), asbestos, talc
VIRGINIA		Tobacco products, chemicals, food products, electric equipment, transportation equipment	Milk, beef cattle, tobacco, broilers, soybeans, clams, crabs, menhaden, oysters	Coal, stone, lime, sand and gravel
WASHINGTON		Transportation equipment, lumber and wood products, food products, metals, paper products, nonelectric machinery	Wheat, beef cattle, milk, apples, potatoes, forest products, salmon, oysters, crabs, tuna	Coal, sand and gravel, stone, gold, gemstones
WEST VIRGINIA		Chemicals, metals, stone and glass and clay products, metal products, electric machinery	Beef cattle, milk, apples, turkeys	Coal, natural gas, petroleum, stone, sand and gravel
WISCONSIN		Nonelectric machinery, food products (especially butter and cheese), paper products, transportation equipment	Milk, beef cattle, hogs, corn	Sand and gravel, stone
WYOMING		Petroleum and coal products, chemicals, clay and glass products, food products	Beef cattle, sheep, sugar beets, wheat, hay, barley	Petroleum, coal, uranium, natural gas, clays

The Trans-Alaska pipeline is 800 miles (1,300 kilometers) long. It carries oil from Prudhoe Bay to the port of Valdez.

THE DOMINION OF CANADA

O Canada!

Our home and native land!

True patriot love

in all thy sons command.

With glowing hearts

we see thee rise,

The True North

strong and free!

From far and wide,

O Canada,

We stand on guard

for thee.

God keep our land

glorious and free!

O Canada,

we stand on guard for thee.

O Canada,

we stand on guard for thee.

"O Canada" was proclaimed Canada's national anthem on July 1, 1980, 100 years after it was first sung. The music was composed by Calixa Lavallée; French lyrics to accompany the music were written by Sir Adolphe-Basil Routhier. Many English versions have appeared over the years, but the version on which the official English lyrics are based was written in 1908 by Mr. Justice Robert Stanley Weir. The official English version includes changes recommended in 1968 by a Special Joint Committee of the Senate and House of Commons.

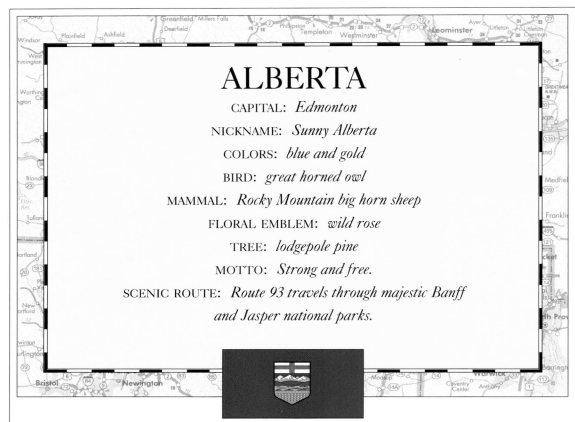

ALBERTA

CAPITAL: *Edmonton*

NICKNAME: *Sunny Alberta*

COLORS: *blue and gold*

BIRD: *great horned owl*

MAMMAL: *Rocky Mountain big horn sheep*

FLORAL EMBLEM: *wild rose*

TREE: *lodgepole pine*

MOTTO: *Strong and free.*

SCENIC ROUTE: *Route 93 travels through majestic Banff and Jasper national parks.*

ALBERTA is Canada's fourth-largest province in area and population. It was named for Princess Louise Caroline Alberta, one of Queen Victoria's daughters and the wife of a Canadian governor general. Because of its unusually good weather, the province is nick-named "Sunny Alberta." It is an area rich in natural resources and beautiful scenery.

The Alberta Plain, in the central part of the province, is a prime farming region. This area's rich soil and regular rainfall help make Alberta a leading producer of barley, oats, and beef cattle. The province is also one of the most oil-rich areas in North America. Alberta supplies most of Canada's petroleum and 80 percent of its natural gas. More than half of Canada's known coal deposits are also located in Alberta.

The towering Canadian Rocky Mountains, along the province's southwestern border, are a popular vacation area. Each year, millions of people travel there to ski, hike, climb, boat, swim, and fish in the Banff, Jasper, and Waterton Lakes national parks. Their snow-covered peaks and picturesque valleys offer spectacular sights. Icy, turquoise Lake Louise, also named after Princess Alberta, in Banff National Park, is a particularly beautiful spot. At the Columbia Icefield in Banff and Jasper, huge glaciers remain from the great ice sheet that once covered most of Canada. Runoff from the glaciers feeds many major river systems in North America. Visitors enjoy riding in snow-mobiles over magnificent Athabasca Glacier.

Calgary, the site of the 1988 Winter Olympics, is Alberta's largest city. The provincial capital, Edmonton, is located on the North Saskatchewan River in the central part of the province and is its second-largest city.

Dinosaur Provincial Park, in eastern Alberta near Brooks, has one of the richest fossil beds in the world. Dinosaur skeletons and fossil remains from 75 million years ago can be seen in the badlands and prairie regions along the Red Deer River.

For more information about Alberta call (800)661-8888

Left, aerial view of the Alberta Plain; *far left,* Maligre Lake from the Bald Hills, Jasper National Park; *top left,* Athabasca Glacier, Jasper National Park; *top,* Lake Louise, Banff; *opposite page,* Edmonton in winter.

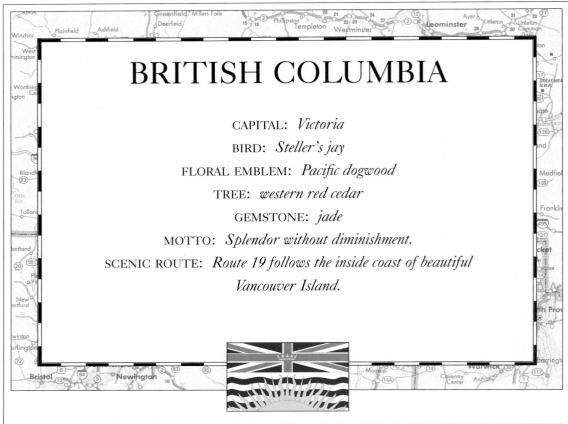

BRITISH COLUMBIA

CAPITAL: *Victoria*

BIRD: *Steller's jay*

FLORAL EMBLEM: *Pacific dogwood*

TREE: *western red cedar*

GEMSTONE: *jade*

MOTTO: *Splendor without diminishment.*

SCENIC ROUTE: *Route 19 follows the inside coast of beautiful Vancouver Island.*

BRITISH COLUMBIA is Canada's Pacific Coast province. Third-largest province in size, it also includes the Queen Charlotte Islands and 285-mile-long (459-kilometer-long) Vancouver Island off the mainland. Along the jagged coastline, fishing is an important industry, with large catches of salmon, halibut, coho, clam, and oyster. The modern city of Vancouver is British Columbia's largest city and an important port on the mainland. It is a key center for trade and is nicknamed "Canada's Doorway to the Orient."

The islands along British Columbia's coast protect ships from the open sea. They offer a scenic, sheltered route from Seattle, Washington, through Vancouver's beautiful Horseshoe Bay, and north to Juneau, Alaska. This path is known as the "Inside Passage to Alaska," and is used by cargo and cruise ships.

British Columbia's highest concentration of population is found in the southwestern part of the province and on Vancouver Island. There, the California Current brings abundant rainfall and a mild climate. Victoria, located on the southeastern end of Vancouver Island, is the provincial capital. With its stone government buildings, tearooms, double-decker buses, and prim English gardens, it is quintessentially British. Some people argue that this charming city is even more British than Britain!

Early Native American tribes along the coast were skilled fishermen and whalers. They carved totem poles and masks, and lived in rectangular houses built from cedar. Thunderbird Park in Victoria features one of the finest collections of totem poles in the world.

The interior of British Columbia is a plateau cut by a series of parallel mountain ranges running northwest to southeast across the province. Winters in these rugged regions are quite severe. These sparsely populated areas are known for their spectacular scenery.

Lumber and the manufacture of wood and paper products are important industries. British Columbia provides Canada with three-fifths of its lumber supply. The Rocky Mountain Trench, a long valley west of the Rockies, is the source of the mighty Fraser and Columbia rivers. British Columbia's great geographical diversity offers homes to more species of birds and mammals than any other province.

For more information about British Columbia call (800) 663-6000

Clockwise from lower left, Gastown Steam Clock, Vancouver; visiting yachts in a Victoria harbor; the Legislative Building, Victoria; K'san Indian Village, near Hazelton; Kootenay River, Kootenay National Park; *top,* Buchart Gardens, Victoria.

MANITOBA is the easternmost of the three prairie provinces and is located in the heart of Canada. Because of its location in the center, or keystone, of the formation made by the 10 provinces, it is nicknamed the "Keystone Province."

The Red River Valley in southern Manitoba is a rich farming area. Wheat, barley, canola, and sunflower are primary crops. Manitoba's capital and largest city is Winnipeg, located at the confluence of the Red and Assiniboine rivers. Because of its strategic location near waterways, highways, and railroads, it is an important industrial and transportation center. Archeological studies indicate that this spot has been a hub of civilization for more than 6,000 years. Winnipeg developed around Fort Garry, one of the Hudson Bay Company's chief fur-trading centers in the 1800s. Today, Lower Fort Garry still stands intact and is a national historic park.

The northern two-thirds of Manitoba is part of the Canadian Shield, a rocky, sparsely-populated area with many rivers and glacial lakes and important deposits of nickel, copper, and gold. Manitoba has more than 100,000 lakes. Three of them—Lake Winnipeg, Lake Manitoba, and Lake Winnipegosis—are so large that they are called the "Great Lakes of Manitoba." Boating, swimming, fishing, and hunting are popular activities.

The name *Manitoba* was likely derived from the Algonquin words *manito waba*, meaning "great spirit's strait." Early Native Americans believed that the noises of the waves in a strait of Lake Manitoba were made by *manito*, the great spirit.

Manitoba has 400 miles (645 kilometers) of coastline on the Hudson Bay. Churchill, on the shore of the bay, is Canada's northernmost sub-arctic port. This area has the greatest concentration of polar bears found anywhere in the world. The bears spend winter hunting for seals on the frozen ice of Hudson Bay and come ashore in the summer. Hudson Bay is also home to many beluga whales.

Many people visit the International Peace Garden on the border of Manitoba and North Dakota. The beautiful plantings there are a symbol of the longstanding friendship between Canada and the United States.

For more information about Manitoba call (800) 665-0040

Top, a member of Manitoba's great polar bear population; *clockwise from lower left,* Northern Boreal Forest; Fort Garry; harvest time in Brandon; the Legislative Building, Winnipeg.

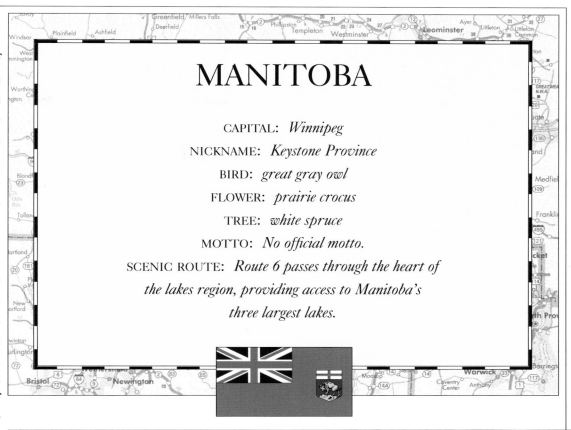

MANITOBA

CAPITAL: *Winnipeg*

NICKNAME: *Keystone Province*

BIRD: *great gray owl*

FLOWER: *prairie crocus*

TREE: *white spruce*

MOTTO: *No official motto.*

SCENIC ROUTE: *Route 6 passes through the heart of the lakes region, providing access to Manitoba's three largest lakes.*

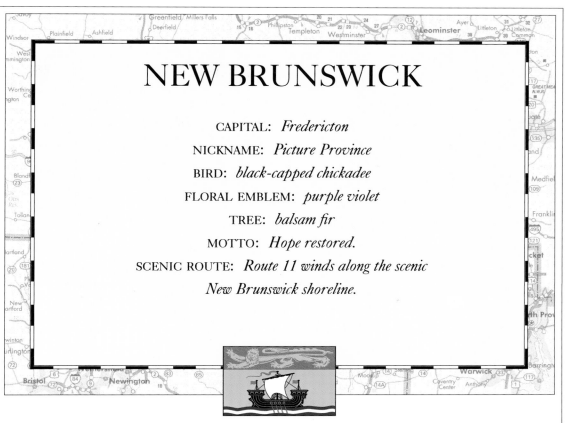

NEW BRUNSWICK

CAPITAL: *Fredericton*

NICKNAME: *Picture Province*

BIRD: *black-capped chickadee*

FLORAL EMBLEM: *purple violet*

TREE: *balsam fir*

MOTTO: *Hope restored.*

SCENIC ROUTE: *Route 11 winds along the scenic New Brunswick shoreline.*

NEW BRUNSWICK is surrounded by water on three sides and has 1,400 miles (2,253 kilometers) of coastline. The tides in the Bay of Fundy, south of New Brunswick, are among the highest in the world. The bay's funnel shape causes the waters to rise as much as 52 feet (16 meters)! Numerous islands in the bay are known for their beautiful beaches. For years prior to his presidency, Franklin Delano Roosevelt had a summer home on Campobello Island.

About 90 percent of New Brunswick is forested. The manufacturing of wood and paper products is the province's biggest industry. Because of its beautiful woods and scenic shorelines, New Brunswick is nicknamed the "Picture Province." The St. John and Miramichi rivers offer excellent sites for camping, hunting, and fishing. New Brunswick is considered one of the top fishing spots in North America.

In 1784, the province was named in honor of England's King George III, the Duke of Brunswick. Many of New Brunswick's early settlers were American colonists who had remained loyal to England. During and after the American Revolution, thousands left America and resettled in New Brunswick. Because of the influence of these Loyalists, New Brunswick is also known as the "Loyalist Province." Today, a recreated village built to show how the Loyalists lived in the 1800s is found at Kings Landing Historical Settlement in Prince William.

New Brunswick's largest city, St. John, is an important center of industry and shipping. Because the harbor does not freeze during the winter, St. John is a busy seaport year-round. The provincial capital, Fredericton, is on the banks of the St. John River. In nearby Hartland, visitors can travel on the longest covered bridge in the world. This bridge spans the St. John River and measures 1,282 feet (391 meters).

Another popular attraction is Magnetic Hill near Moncton. If a car is put in neutral gear at the bottom of the hill, it appears to defy gravity and roll up to the top of the hill! This is really an optical illusion.

For more information about New Brunswick call (800) 561-0123

Top, salmon fishing in the Miramichi River; *clockwise from lower left,* the longest covered bridge in the world, Hartland; sea kayaking along Hole in the Wall, Grand Manan Island; St. John; the historical settlement of Kings Landing; *opposite page,* Point-du-Chene, Northumberland Strait.

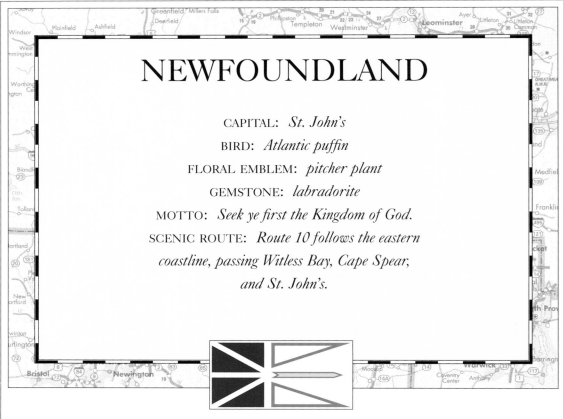

NEWFOUNDLAND

CAPITAL: *St. John's*

BIRD: *Atlantic puffin*

FLORAL EMBLEM: *pitcher plant*

GEMSTONE: *labradorite*

MOTTO: *Seek ye first the Kingdom of God.*

SCENIC ROUTE: *Route 10 follows the eastern coastline, passing Witless Bay, Cape Spear, and St. John's.*

NEWFOUNDLAND, Canada's newest province, was established in 1949. Its boundaries include Labrador on the Canadian mainland and the island of Newfoundland. From the rocky cliffs of the island's Cape Spear, south of St. John's, visitors can be the first to see the sun rise in North America.

Newfoundland is more than three times larger than the other three maritime provinces combined, but it has fewer people than any province except Prince Edward Island. Most of the population lives along the rocky coasts of Newfoundland Island in small fishing villages. For hundreds of years, the surrounding waters have been prime fishing grounds for crews from around the world. Labrador is part of the Canadian Shield, a rugged plateau of ancient rocks. This area is heavily forested, with many large lakes and rivers. Caribou, moose, and black bear are plentiful.

Five hundred years before Columbus set sail, Norse Vikings settled in L'Anse aux Meadows on the northern tip of the island of Newfoundland. Today a recreated Viking village, with turf-walled buildings, can be seen. Historians believe John Cabot sailed into the harbor of present-day St. John's in 1497 and claimed the *new found land* for England. It is also believed that Cabot named the site after Saint John the Baptist because he sailed there on June 24, the saint's feast day. St. John's is one of the oldest settlements in North America and is Newfoundland's capital and largest city. In 1901, Italian inventor Guglielmo Marconi received the first transatlantic wireless signal on Signal Hill in St. John's.

Newfoundland's climate is generally cool due to its Arctic winds and sea currents. Skiing and snowmobiling are popular activities. Corner Brook, Newfoundland's only other major city, receives about 16 feet (4.8 meters) of snow each year—more than any other city in Canada!

Bird lovers travel to Newfoundland to see its more than 300 different species of birds. The islands in Witless Bay house the largest Atlantic puffin colony in North America. More than 22 species of whale and dolphin also live along the coast.

For more information about Newfoundland call (800) 563-6353

Left, Burnside reenactment at Cabot Tower, St. John's; *middle left,* Western Brook Pond, Gros Morne National Park; *top left,* Cape Spear; *top,* an Atlantic puffin.

84

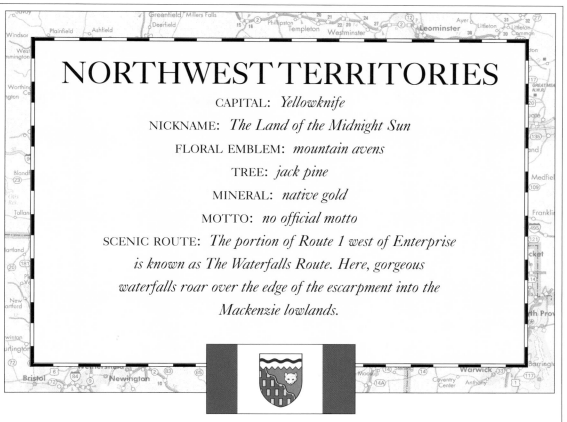

THE Northwest Territories cover about one-third of Canada, yet house less than one percent of the country's population. Half of this vast area is north of the Arctic Circle and stretches within 500 miles (800 kilometers) of the North Pole. The Northwest Territories have an incredible variety of geographical features and natural resources. Large deposits of zinc, lead, iron ore, gold, and petroleum make mining the area's top industry. Fur trapping is also important.

The tree line divides the mainland almost diagonally from the Mackenzie River Delta in the northwest to a point on the Hudson Bay near the Manitoba border. South and west of this line are mountain ranges, grasslands, and large forests. Wildlife species include grizzly and black bear, moose, caribou, Dall sheep, and mountain goat. Summers are short and mild, but winters are long and challenging. The Mackenzie River, Canada's longest at 1,071 miles (1,724 kilometers), is frozen for eight to nine months each year. Yellowknife, named after the copper knives carried by the Native Americans, is the capital and largest city. Located on the northern shore of Great Slave Lake, it is a base for exploring the rest of the Territories. From there, planes regularly fly travelers to more remote areas.

North of the tree line, the summers are short and cool, with an average July temperature of 42 degrees Fahrenheit (6 degrees Celsius) along the coast of Hudson Bay. All of this area is permafrost, perennially frozen ground or gravel. Little can grow there. Many Eskimos, or Inuits, and other native peoples live along the Hudson Bay and on nearby Baffin Island, one of the largest islands in the world. Farther north, conditions are even more extreme. Average January temperatures on the northern parts of the mainland and the Arctic islands are around minus 40 degrees Fahrenheit (minus 40 degrees Celsius). Jagged mountains, huge glaciers, and deep fjords offer breathtaking scenery for the small group of Inuits and few government employees who live and work there. Polar bear, seal, walrus, arctic fox, caribou, and whale inhabit this region.

Of the Territories' four national parks, only Wood Buffalo is accessible by road. The others are remote wilderness areas. Organized development of the Northwest Territories did not begin until the 1950s. Today, much remains to be discovered and explored.

For more information about the Northwest Territories call (800) 661-0788

Top, an Inuk girl wearing a caribou parka; *clockwise from lower left,* Yellowknife; a dog team; an iceberg; hiking in Auyuittuq National Park, Baffin Island.

NORTHWEST TERRITORIES

CAPITAL: *Yellowknife*

NICKNAME: *The Land of the Midnight Sun*

FLORAL EMBLEM: *mountain avens*

TREE: *jack pine*

MINERAL: *native gold*

MOTTO: *no official motto*

SCENIC ROUTE: *The portion of Route 1 west of Enterprise is known as The Waterfalls Route. Here, gorgeous waterfalls roar over the edge of the escarpment into the Mackenzie lowlands.*

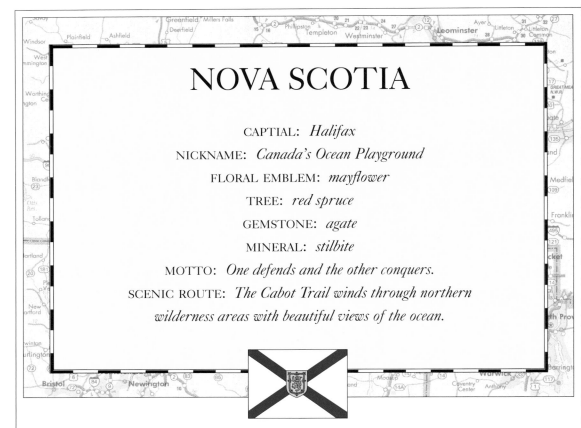

NOVA SCOTIA

CAPTIAL: *Halifax*

NICKNAME: *Canada's Ocean Playground*

FLORAL EMBLEM: *mayflower*

TREE: *red spruce*

GEMSTONE: *agate*

MINERAL: *stilbite*

MOTTO: *One defends and the other conquers.*

SCENIC ROUTE: *The Cabot Trail winds through northern wilderness areas with beautiful views of the ocean.*

NOVA SCOTIA is almost completely surrounded by water. This province includes Cape Breton Island and a mainland peninsula that reaches out into the Atlantic Ocean. Although Nova Scotia is the second-smallest province, its irregular coastline measures 4,709 miles (7,578 kilometers). With its pretty beaches and rocky coasts, Nova Scotia attracts many visitors each year and is referred to as "Canada's Ocean Playground." The surrounding ocean and bay waters keep its climate from being extremely cold or hot. These waters also yield large quantities of lobster, cod, scallop, and other seafood. Further inland, salmon, shad, and trout are plentiful. Nova Scotia's fishing industry is ranked first in Canada.

The Bay of Fundy separates most of Nova Scotia from New Brunswick. The narrow Strait of Canso splits the Nova Scotia mainland from Cape Breton Island. A large saltwater lake called Bras d'Or almost splits the island in two. About 3,800 small islands, most of them uninhabited, sit off the province's coast. The largest of these is Sable Island, a long sandbar that has been the site of countless shipwrecks.

Almost 80 percent of Nova Scotia is forested. Camping and hiking are popular activities in the province's parks. Paper and wood pulp manufacturing is an important industry.

Alexander Graham Bell, teacher of the deaf and inventor of the telephone, spent his later years in Nova Scotia. Visitors to the Alexander Graham Bell National Historic Park, near Baddeck on Cape Breton Island, can see his models, inventions, and photographs.

British settlers from Scotland arrived in Nova Scotia in 1629. They named the new land *Nova Scotia,* Latin for "New Scotland." The Scottish influence is still seen today. Many towns hold celebrations complete with authentic Scottish clothing, songs, and bagpipe music. Some people still speak the Gaelic language of their ancestors. Nova Scotia was one of the four original provinces. In 1867, it united with New Brunswick, Ontario, and Quebec to form the Dominion of Canada.

Halifax, the provincial capital, is located on Halifax Harbor and is an important center for international trade. Two suspension bridges span the harbor to link Halifax with Dartmouth, the province's second-largest city.

For more information about Nova Scotia call (902) 424-4247

Clockwise from lower left, Neils Harbor; Peggy's Cove; Halifax; whale watching; *top,* rock formations on the shores of the Minas Basin, Annapolis Valley; *opposite page,* Cabot Trail, Cape Breton Island.

ONTARIO

CAPITAL: *Toronto*

NATIONAL CAPITAL: *Ottawa*

BIRD: *common loon*

FLORAL EMBLEM: *white trillium*

GEMSTONE: *amethyst*

TREE: *white pine*

MOTTO: *Loyal she began, loyal she remains.*

SCENIC ROUTE: *The 205-mile Lake Nipissing Circle Route
passes historic small towns, spectacular lookout points,
and the thriving city of North Bay.*

ONTARIO is the second-largest province and ranks first in population, housing one-third of Canada's people. Ninety percent of Ontario's residents live in the southeastern part of the province, south of Lake Nipissing. Toronto, the provincial capital, and Ottawa, Canada's capital, are in this region.

Ontario, the "Manufacturing Heartland of Canada," produces nearly half of Canada's manufactured goods and provides one-fourth of the country's farm products. Ontario's top industry is automobile manufacturing. Situated along four of the five Great Lakes, and with easy access to European markets via the St. Lawrence Seaway, Ontario has a strategic location on the North American continent.

Along with Quebec, New Brunswick, and Nova Scotia, Ontario is one of Canada's four original provinces. In 1857, Queen Victoria chose Ottawa, then a lumber village, to be the capital of Canada. Today, Ottawa is a graceful, cosmopolitan city. Its stately stone Parliament buildings stand on a bluff overlooking the Ottawa River.

Toronto, a major port on Lake Ontario, has the largest metropolitan area population of any city in Canada. It is the hub of Canada's industry and finance. With its modern architecture and many museums and theaters, Toronto is also a cultural center for the country.

Rivers, lakes, and waterfalls cover one-sixth of Ontario. Throughout the province, fishing, boating, swimming, and camping are popular activities.

Each year, millions of tourists travel to see spectacular Niagara Falls, on the border between the United States and Canada. More Kodak film is sold there than anywhere else in the world! Horseshoe Falls, on the Canadian side, is the largest part of Niagara Falls. The Niagara River is one of the shortest, and wildest, rivers in the world. Its rapids reach speeds of 30 miles per hour (48 kilometers per hour).

The name *Ontario* is taken from an Iroquois word, possibly meaning "rocks standing high" or "near the water," referring to Niagara Falls.

More Native Americans live in Ontario than in any other province. Most of them live on reservations in the northwest. This rocky area is part of the Canadian Shield and has rich deposits of nickel, gold, uranium, and amethyst.

For more information about Ontario call (800) ONTARIO (688-2746)

Clockwise from lower left, Horseshoe Falls, the Canadian side of Niagara Falls; the Parliament Buildings, Ottawa; Toronto City Hall; the Sky Dome, Toronto; *top,* Toronto.

PRINCE EDWARD ISLAND, in the Gulf of St. Lawrence, is the smallest of Canada's provinces and has the fewest number of people. Because of its small size, however, it is the most densely populated. The island is a fertile plain of rich, red soil. Much of the land is used for farming, with potatoes the primary crop. Prince Edward Island is nicknamed the "Garden of the Gulf" and "Million Acre Farm."

Along the island's 500-mile (805-kilometer) coastline, fishing is an important industry. Prince Edward Island is a leader in oyster and lobster production. The island's warm coastal waters give it a much milder climate than is found on the Canadian mainland. Swimming, sailing, fishing, golfing, and biking are popular activities there.

Malpeque and Hillsborough bays nearly divide Prince Edward Island into three parts. About 40 towns and villages are scattered throughout the island. Charlottetown is the provincial capital and the island's only city. In 1864, Canada's founding fathers gathered in Province House in Charlottetown and planned the Confederation of Canada. Confederation Chamber there is known as the "Birthplace of Canada."

In her book, *Anne of Green Gables*, Lucy Maud Montgomery affectionately described life on Prince Edward Island. The Green Gables farmhouse from her story is found in Prince Edward Island National Park, along a fairway of Green Gables Golf Course, which is one of Canada's best courses. Cavendish and Rainbow Valley are other island sites featured in Montgomery's subsequent books.

Prince Edward Island's earliest peoples were the Micmac Indians. In the 1700s, the island became a British colony and was named for Queen Victoria's father, Edward, Duke of Kent. Today visitors can tour Micmac Village, near Rocky Point, to see a recreated Native American settlement as it would have looked in the 1700s. Another popular tourist site is Woodleigh Replicas near Burlington. Over a period of 30 years, large-scale models of famous British homes, castles, and churches were built in an English garden setting. Some of the detailed replicas are large enough to enter.

For more information about Prince Edward Island call (800) 463-4PEI (463-4734)

PRINCE EDWARD ISLAND

CAPITAL: *Charlottetown*

NICKNAME: *Garden of the Gulf*

BIRD: *blue jay*

FLORAL EMBLEM: *lady's-slipper*

MOTTO: *The small under the protection of the great.*

SCENIC ROUTE *The Blue Heron Scenic Drive, in the central part of the island, winds through PEI National Park and the capital of Charlottetown.*

Top, Woodleigh replicas, Burlington; *right,* Green Gables farmhouse, Prince Edward Island National Park; *top right,* Cape Kildare.

QUÉBEC is Canada's largest province and ranks second in population behind Ontario. Whereas the rest of Canada is influenced by Great Britain, 80 percent of Québec's people are of French ancestry and follow French customs and traditions. A large percentage of French Canadians speak French as their first language, and most of Québec's schools use it as the language of instruction. Throughout the province, this French influence gives Québec a unique atmosphere.

The St. Lawrence River, which connects the Atlantic Ocean with the Great Lakes, has been an important trade route since French explorer Jacques Cartier sailed it in 1535. It is often called the "Mother of Canada." The fertile land along this river in southern Québec is one of the most populated regions in Canada. Québec's capital city, also named Québec, and its largest city, Montréal, are located there.

In 1608, the French explorer Samuel deChamplain founded Québec City on the bluffs overlooking the point where the St. Charles and St. Lawrence rivers converge. The St. Lawrence River is only about one-half mile (0.8 kilometers) wide near this point. The name *Québec* was taken from the Algonquian Indian word *kebec*, meaning "the place where the river narrows." As the oldest city in Canada, Québec is rich in history and charm. The original city environs are walled, and several sections have houses and churches dating from the 1600s and 1700s. In 1759, British troops scaled the sheer cliffs by the city and defeated the French, winning possession of the province. But more than 200 years later, Québec is English in name only. Its prime location makes it an important shipping and manufacturing center.

Among French-speaking cities of the world, Montréal ranks second in population behind Paris. It is located on the island of Montréal at the juncture of the St. Lawrence and Ottawa rivers. Montréal was named for Mount Royal (*Mont Réal* in French), a tree-covered mountain that stands in the center of the city.

Nine-tenths of Québec is part of the Canadian Shield, an ancient, rugged, rocky plateau. Deposits of iron ore, asbestos, copper, gold, and zinc are found there. Large forests cover half the province. Québec is Canada's top paper-producing province and ranks first in North America for maple syrup production.

For more information about Québec call (418)643-1344

Top, a sidewalk cafe in Montréal; *right,* Montréal; *far right,* L'Anse St. Jean, Saguenay River; *top right,* Cabin au Sucre, St. Jean de-Brebeuf; *opposite page,* Chateau Frontenac, Québec City.

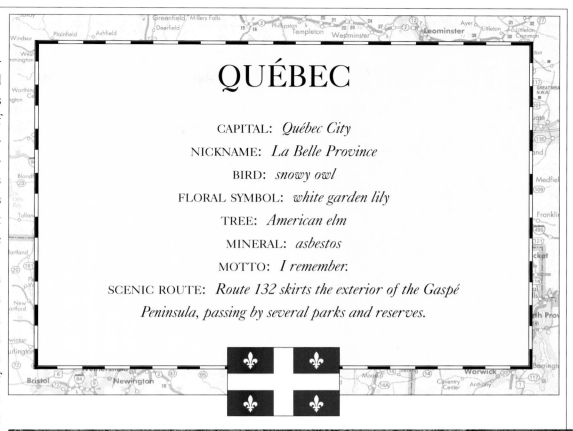

QUÉBEC

CAPITAL: *Québec City*

NICKNAME: *La Belle Province*

BIRD: *snowy owl*

FLORAL SYMBOL: *white garden lily*

TREE: *American elm*

MINERAL: *asbestos*

MOTTO: *I remember.*

SCENIC ROUTE: *Route 132 skirts the exterior of the Gaspé Peninsula, passing by several parks and reserves.*

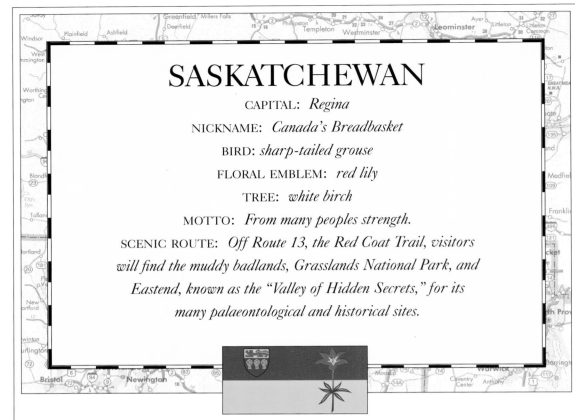

SASKATCHEWAN

CAPITAL: *Regina*

NICKNAME: *Canada's Breadbasket*

BIRD: *sharp-tailed grouse*

FLORAL EMBLEM: *red lily*

TREE: *white birch*

MOTTO: *From many peoples strength.*

SCENIC ROUTE: *Off Route 13, the Red Coat Trail, visitors will find the muddy badlands, Grasslands National Park, and Eastend, known as the "Valley of Hidden Secrets," for its many palaeontological and historical sites.*

SASKATCHEWAN is named for the great Saskatchewan River that flows across the province. Cree Indians called the river *Kisiskatchewan*, which means "swift-flowing." Saskatchewan is one of the prairie provinces. The vast, fertile plains in the southern part of the province are well suited for farming and raising beef cattle. Saskatchewan is the top-producing wheat region in North America and supplies about half of Canada's crop. With high yields of barley and rye, too, the province earns the nickname "Canada's Breadbasket." Rich supplies of oil make Saskatchewan a leading oil producer for North America.

Most of Saskatchewan's population lives in the southern two-thirds of the province. The northern part of Saskatchewan is a rugged, heavily wooded area with many large glacial lakes. This region is sparsely populated. Fishing enthusiasts travel there for the abundant supplies of grayling, trout, and northern pike. Connecting one lake to another, the Churchill River offers excellent canoeing and whitewater rafting. Hunting is another popular form of recreation. Each fall, three major migratory flyways converge over Saskatchewan, bringing huge flocks of geese. The province is also a breeding ground for about one-third of the duck population of North America. Northern Saskatchewan has many caribou, moose, elk, and bear.

Saskatchewan has short, warm summers and long, cold winters, but temperatures vary greatly throughout the province. Average January temperatures in the south are 10 degrees Fahrenheit (minus 12 degrees Celsius) as compared to minus 23 degrees Fahrenheit (minus 31 degrees Celsius) in the north.

Regina, Saskatchewan's capital and largest city, was named in honor of Queen Victoria. The training headquarters for the Royal Canadian Mounted Police are located there. In Saskatoon, the province's second-largest city, visitors can tour the Western Development Museum and view a re-created frontier town of the early 1900s.

For more information about Saskatchewan call (800) 667-7191

Left, a rodeo; *far left,* a deer, one of the many abundant forms of wildlife found in Saskatchewan; *middle left,* Lefty Falls; *top left,* wheat fields in "Canada's Breadbasket"; *top,* Clearwater River.

IN 1898, at the height of the Klondike Gold Rush, the Yukon became a federal territory separate from the Northwest Territories. It was named after the Yukon River, one of the longest rivers in North America. Native Americans in the area called it *Yu-kun-ah* which meant "greatest" or "big river."

The Yukon Territory is shaped somewhat like a triangle and reaches from Alaska east to the Northwest Territories and from British Columbia north to the Beaufort Sea. Much of the terrain is quite rugged. Several mountain ranges traverse the territory. Mt. Logan, in the St. Elias Mountains, is the highest point in Canada at 19,590 feet (5,971 meters).

When gold was discovered in Bonanza Creek, a tributary of the Klondike River, thousands of hopefuls flocked to the area. At the peak of the gold rush, more people lived in the Yukon than live there now, and a lucky few made their fortunes. In 1900, hand dredging alone produced gold worth more than $22 million! Today, in addition to gold, important deposits of zinc, lead, silver, coal, asbestos, and nickel are found there.

More than half of the Yukon Territory is forested. Fishing, canoeing, cross-country skiing, and hunting are popular activities. Resident wildlife includes caribou, elk, mountain goat, Dall sheep, moose, grizzly and black bear, and many game birds. The Alaska and Dempster highways pass through the Yukon and offer scenic views of unspoiled wilderness.

Whitehorse, on the Yukon River in the south-central part of the territory, is the capital and only city. Two-thirds of the territory's people live there. Whitehorse is the transportation, distribution, and communication center for the entire territory and the territorial headquarters for the Royal Canadian Mounted Police.

Summers in the Yukon are short and mild, but winters are long and harsh. Throughout the territory, temperatures vary widely with much colder weather farther north above the Arctic Circle. The coldest temperature ever recorded in North America was a reading of minus 81 degrees Fahrenheit (minus 63 degrees Celsius) at Snag Airport along the Alaskan border.

For more information about the Yukon Territory call (403) 667-5340

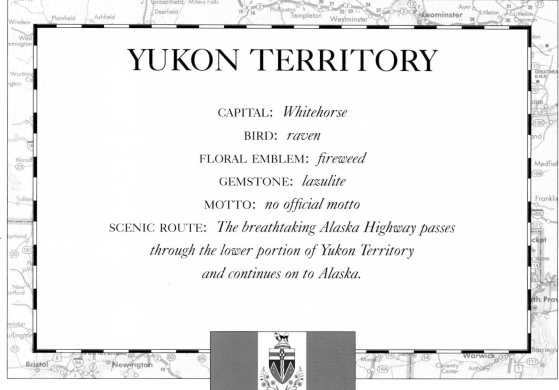

YUKON TERRITORY

CAPITAL: *Whitehorse*

BIRD: *raven*

FLORAL EMBLEM: *fireweed*

GEMSTONE: *lazulite*

MOTTO: *no official motto*

SCENIC ROUTE: *The breathtaking Alaska Highway passes through the lower portion of Yukon Territory and continues on to Alaska.*

Top, building an igloo; *right,* the Yukon wilderness; *top right,* North Canol Road Rapids on south MacMillan River with the MacKenzie mountain range in the distance.

RANKING OF THE PROVINCES AND TERRITORIES BY AREA

	PROVINCES AND TERRITORIES	SQUARE MILES	SQUARE KILOMETERS
1	NORTHWEST TERRITORIES	1,322,900	3,379,684
2	QUEBEC	594,860	1,540,680
3	ONTARIO	412,582	1,068,582
4	BRITISH COLUMBIA	366,255	948,596
5	ALBERTA	255,285	661,185
6	SASKATCHEWAN	251,700	652,900
7	MANITOBA	250,947	650,087
8	YUKON TERRITORY	186,300	482,515
9	NEWFOUNDLAND	156,185	404,517
10	NEW BRUNSWICK	28,354	73,437
11	NOVA SCOTIA	21,425	55,490
12	PRINCE EDWARD ISLAND	2,185	5,657

RANKING OF THE PROVINCES AND TERRITORIES BY POPULATION*

1	ONTARIO	10,928,000
2	QUEBEC	7,281,000
3	BRITISH COLUMBIA	2,668,000
4	ALBERTA	2,716,000
5	MANITOBA	1,131,000
6	SASKATCHEWAN	1,016,000
7	NOVA SCOTIA	937,000
8	NEW BRUNSWICK	759,000
9	NEWFOUNDLAND	582,000
10	PRINCE EDWARD ISLAND	135,000
11	NORTHWEST TERRITORIES	64,000
12	YUKON TERRITORY	30,000

*BASED ON 1994 POST-CENSUS ESTIMATES.

CHRONOLOGICAL LISTING OF PROVINCES AND TERRITORIES ENTERING THE CONFEDERATION

NEW BRUNSWICK	JULY 1, 1867
NOVA SCOTIA	JULY 1, 1867
ONTARIO	JULY 1, 1867
PRINCE EDWARD ISLAND	JULY 1, 1867
QUEBEC	JULY 1, 1867
MANITOBA	JULY 15, 1870
NORTHWEST TERRITORIES	1870
BRITISH COLUMBIA	JULY 20, 1871
YUKON TERRITORY	1898
ALBERTA	1905
SASKATCHEWAN	SEPTEMBER 1, 1905
NEWFOUNDLAND	MARCH 31, 1949

FASCINATING FACTS ABOUT THE PROVINCES AND TERRITORIES

The West Edmonton Mall, in Alberta, features an N.H.L-size hockey rink, an indoor water park, an indoor amusement park, submarine rides, and a miniature golf course—in addition to more than 700 stores.

Over $95 million in gold was mined from the Klondike region between 1896 and 1903.

The largest non-polar ice field in the world is located in the St. Elias mountain ranges. It's believed to be 2,297 feet (700 meters) deep in the heart of the mountains.

North America's largest population of grizzly bear and Dall sheep live in the Yukon.

The largest freshwater bar dune system in the world is found at Sandbanks Provincial Park on the shores of Lake Ontario, near Picton.

The only freshwater underwater park in North America, Fathom Five National Park, allows visitors to view nineteenth century shipwrecks either by scuba diving or touring on a glass-bottom boat.

The C.N. Tower in Toronto is the tallest freestanding structure in the world. It is over 1,815 feet (553.3 meters) from the pool at its base to the tip of the transmission mast at the top—twice as tall as the Eiffel Tower in France.

The Toronto Skydome was the first stadium in the world with a fully retractable roof.

North America's oldest bird sanctuary, founded in 1887, is at Last Mountain Lake.

Once a year, great white polar bear roam freely through the streets of Churchill, Manitoba, until ice forms on the Hudson Bay and enables them to hunt for seal.

Québec is the only walled city north of Mexico.

NORTHERNMOST POINT: Cape Columbia on Ellesmere Island, Northwest Territories

SOUTHERNMOST POINT: Middle Island in Lake Erie, Ontario

EASTERNMOST POINT: Cape Spear, Newfoundland

WESTERNMOST POINT: St. Elias Mountains, Yukon Territory

HIGHEST POINT: Mount Logan in the St. Elias Mountains in the Yukon Territory; 19,520 feet (5,950 meters) tall. This is the second-highest peak in North America.

HIGHEST TIDES: The Bay of Fundy between Nova Scotia and New Brunswick; 52 feet (16 meters) or more. The Fundy tides are the highest in the world.

LARGEST LAKE: Great Near Lake in the Northwest Territories; 12,175 square miles (31,792 square kilometers)

LONGEST RIVER: Mackenzie River; 1,071 miles (1,724 kilometers) long

LARGEST ISLAND: Baffin Island in the Northwest Territories; 195,927 square miles (507,449 square kilometers). This is the sixth-largest island in the world.

OLDEST NATIONAL PARK: Banff National Park; established in 1885 in south-western Alberta

MAJOR PRODUCTS AND INDUSTRIES OF THE PROVINCES AND TERRITORIES

	MANUFACTURING	FARMING AND FISHING	MINING
ALBERTA	Food products, chemicals, metal products, machinery, petroleum and coal products	Beef cattle, wheat, rye, barley, oats, hogs, milk	Petroleum, natural gas, coal, sulfur, sand and gravel, salt, clay
BRITISH COLUMBIA	Wood products, paper products, food products, primary metals	Milk, beef cattle, fruits, salmon, halibut, herring, coho, clams, oysters	Coal, copper, natural gas, petroleum
MANITOBA	Food products, machinery, metal products, transportation equipment	Wheat, barley, beef cattle, hogs, oats, canola, sunflowers	Nickel, copper, petroleum, zinc, gold
NEW BRUNSWICK	Paper products, wood products, food products, glass products	Potatoes, milk, poultry, lobster, herring, salmon, bluefin tuna	Zinc, lead, copper, silver, coal
NEWFOUNDLAND	Food products, paper products, chemicals, transportation equipment	Eggs, poultry, blueberries, flounder, lobster, cod, salmon, shrimp, scallops	Iron ore, asbestos, zinc, sand and gravel
NOVA SCOTIA	Food products, paper products, transportation equipment, chemicals, wood products, printed materials, textiles	Milk, hogs, poultry, eggs, beef cattle, fruits, lobster, cod, scallops, haddock, herring, mackerel	Coal, salt, gypsum, limestone, sand and gravel
NORTHWEST TERRITORIES	Food products, petroleum products, wood products	Whitefish, char, northern pike	Zinc, gold, lead, natural gas, petroleum
ONTARIO	Transportation equipment, food products, fabricated metal products	Poultry, eggs, milk, beef cattle, fruit, vegetables, perch, pickerel, whitefish	Nickel, gold, uranium, copper, amethyst
PRINCE EDWARD ISLAND	Food products, printed materials, wood products, furniture	Potatoes, hogs, beef cattle, milk, hay, vegetables, fruits, lobster, cod, oysters, mackerel, bluefin tuna, mussels, scallops	Sand and gravel, natural gas
QUEBEC	Food products, paper products, primary metals, transportation equipment, chemicals, clothing, printed materials, textiles	Milk, hogs, beef cattle, eggs, corn, hay, rye, potatoes, fruits, maple syrup, cod, mackerel, salmon, lobster, crab, shrimp	Iron ore, asbestos, copper, gold, zinc
SASKATCHEWAN	Food products, machinery, printed materials, glass products	Wheat, rye, barley, rapeseed, oats, beef cattle, hogs, poultry, eggs, trout, northern pike, grayling	Petroleum, potash, coal, uranium, sodium sulfate, natural gas, copper, gold, salt
YUKON TERRITORY	Lumber and wood products, printed materials, food products, clothing	Salmon, whitefish	Zinc, lead, gold, silver, coal, asbestos, nickel

FORT McHENRY

Courtesy of Roger Miller, Roger Miller Photo, Ltd.

ALABAMA

State capitol, Cahaba River, and shrimp boats *courtesy of State of Alabama Bureau of Tourism and Travel.* Statue of Vulcan *courtesy of Superstock.*

ALASKA

Eskimo children, caribou, totem carver *(photographer: Nathan Jackson),* and Mendenhall Glacier *courtesy of Alaska Division of Tourism.* Dogsled on the Iditarod Trail *courtesy of Superstock.*

ARIZONA

Grand Canyon, Canyon de Chelly, San Xavier, Tonto Natural Bridge, Havasupai Canyon, and desert sunset *courtesy of Arizona Office of Tourism.* Monument Valley *courtesy of Superstock.*

ARKANSAS

Buffalo National River, the state capitol, and Ozark National Forest *courtesy of A.C. Haralson, Arkansas Department of Parks and Tourism.* The Old Mill *courtesy of Superstock.*

CALIFORNIA

Death Valley, Venice Beach, Santa Barbara, and Heavenly Valley *courtesy of Robert Holmes and the California Office of Tourism.* Gull Rock, San Francisco, and Disneyland *courtesy of Superstock.*

COLORADO

Civic Center Park, Maroon Bells near Aspen, and Denver at night *courtesy of Denver Metro Convention and Visitors Bureau.* Spruce Tree House *courtesy of the National Park Service.* Sylvan Lake State Park *courtesy of Mike Hopper, Colorado State Parks.* Lory State Park and Mencos State Recreational Area *courtesy of Colorado State Parks.*

CONNECTICUT

Courtesy of the Connecticut Department of Economic Development.

DELAWARE

Winterthur Museum, Brandywine Valley, and Rehoboth Beach *courtesy of Delaware Tourism Office, Department of Development.* Hagley Museum *courtesy of Superstock.*

FLORIDA

Miami, Saint Augustine, and Fort Myers *courtesy of Florida Department of Commerce, Division of Tourism.* Flamingos *courtesy of Bruce Smith.* Walt Disney World *courtesy of Superstock.*

GEORGIA

Underground Atlanta, Martin Luther King Jr. memorial, and Okefenokee swamp *courtesy of Georgia Department of Industry, Trade & Tourism.* Stone Mountain Monument, Forsyth Park, and

George Woodruff House *courtesy of Superstock.*

HAWAII

Courtesy of Superstock.

IDAHO

Courtesy of the Idaho Travel Council, Idaho Department of Commerce. Centennial Trail, *Roger Williams;* Silver City hotel, *Steve Bly;* Shoshone Falls, *Bill Grange.*

ILLINOIS

Courtesy of Superstock.

INDIANA

Indiana Dunes, the Indianapolis 500 and the Ohio River *courtesy of the Indiana Department of Commerce, Tourism Division.* State capitol and Dearborn *courtesy of Superstock.*

IOWA

Loess Hills *courtesy of the Iowa Department of Economic Development.* The state capitol and farm *courtesy of Superstock.* .

KANSAS

Courtesy of Superstock.

KENTUCKY

Fort Boonesborough, Cumberland Falls State Park, and bluegrass horse farm *courtesy of the Kentucky Department of Travel Development.* Kentucky farm scene *courtesy of Superstock.*

LOUISIANA

Courtesy of Louisiana Office of Tourism.

MAINE

Mt. Katahdin/Compass Pond and Rockland Harbor *courtesy of the Maine Office of Tourism.* Bass Harbor Lighthouse and Rockport lobsters *courtesy of Superstock.*

MARYLAND

Swallow Falls State Park and the U.S. Naval Academy *courtesy of Maryland Office of Tourism Development.* Inner Harbor and horses *courtesy of Craig Schleunes.*

MASSACHUSETTS

Plimoth Plantation *courtesy of Plimoth Plantation, Inc., Plymouth, Massachusetts USA, Ted Curtin, photographer.* Shutesbury *courtesy of Massachusetts Office of Travel and Tourism, Howard Karger, photographer.* Nauset Lighthouse *courtesy of Massachusetts Office of Travel and Tourism.* Faneuil Hall and Old North Church *courtesy of Greater Boston Convention and Visitors Bureau.* North Truro, Cape Cod *courtesy of Superstock.*

MICHIGAN

Knob Hill and Detroit *courtesy of Michigan Travel Bureau, Don Simonelli, photographer.* Mackinac Bridge, Miner's Castle Rocks and Lake Michigan *courtesy of Michigan Travel Bureau.*

MINNESOTA

Courtesy of Minnesota Office of Tourism.

MISSISSIPPI

Courtesy of Mississippi Division of Tourism.

MISSOURI

Permission granted by the Missouri Division of Tourism.

MONTANA

Courtesy of Travel Montana.

NEBRASKA

Indian Cave State Park *courtesy of Nebraska Department of Economic Development, H.W. Legg, photographer.* Millet, Scotts Bluff National Monument and the Capitol Building *courtesy of Superstock.*

NEVADA

Lake Tahoe, Virginia City, Cathedral Gorge and Hoover Dam *courtesy of Nevada Commission on Tourism.* Excalibur Hotel and Reno *courtesy of Superstock.*

NEW HAMPSHIRE

Marlborough foliage *courtesy of State of New Hampshire Tourism, Paul Spezzaferri, photographer.* Marlow *courtesy of State of New Hampshire Tourism, Arthur Boufford, photographer.* Cannon Mountain *courtesy of State of New Hampshire Tourism, David Brownell, photographer.* Andover *courtesy of Superstock.*

NEW JERSEY

Pine Barrens and Atlantic City beach *courtesy of New Jersey Division of Travel and Tourism.* Kwanzan, Atlantic City at night, and Lamington *courtesy of Superstock.*

NEW MEXICO

Kiowa Grasslands and White Sands National Monument *courtesy of New Mexico Magazine, Mark Nohl, photographer.* Cactus, Balloon Fiesta, and church *courtesy of Superstock.*

NEW YORK

Statue of Liberty and West Point *courtesy of New York State Department of Economic Development.* Treman Park, Niagara Falls, and South Street Seaport *courtesy of Superstock.*

NORTH CAROLINA

Courtesy of Superstock.

NORTH DAKOTA

Courtesy North Dakota Department of Tourism.

OHIO

Courtesy of Ohio Division of Travel and Tourism. Athens, *D. Bugeja, photographer;* Cincinnati at night, *D. Castelli, photographer.*

OKLAHOMA

Courtesy of Oklahoma Tourism and Recreation Department, Fred Marvel, photographer.

OREGON

Courtesy of Superstock.

PENNSYLVANIA

Delaware Water Gap and Valley Forge *courtesy of the Pennsylvania Office of Travel Marketing.* Liberty Bell, Independence Hall, and Amish buggy *courtesy of Superstock.*

RHODE ISLAND

Courtesy of the Rhode Island Tourism Division. Block Island lighthouse: *Browning, photographer.*

SOUTH CAROLINA

Chatooga River, Myrtle Beach, and Hilton Head Island *courtesy of the South Carolina Department of Parks, Recreation and Tourism.* Fort Sumter *courtesy of Bruce Smith.*

SOUTH DAKOTA

Badlands *courtesy of Bruce Smith.* Mount Rushmore *courtesy of the South Dakota Department of Tourism, Mark Kayser, photographer.* Custer State Park *courtesy of the South Dakota Department of Tourism, Chad Coppes, photographer.* Mitchell Corn Palace *courtesy of Superstock.*

TENNESSEE

Courtesy of Tennessee Tourist Development.

TEXAS

Texas panhandle, Rio Grande Valley, west Texas, south Texas dancer, and the gulf coast *courtesy of the Texas Department of Commerce.* Dallas *courtesy of Superstock.*

UTAH

Courtesy of the Utah Travel Council. Rainbow Bridge, John Telford, photographer; Tower Arch, Salt Lake Temple, and Great Salt Lake, *Frank Jensen, photographer;* Bryce Canyon National Park, *Tom Till, photographer;* Sundance Ski Resort, *Kim Despain, photographer.*

VERMONT

East Corinth and Cabot *courtesy of the Vermont Department on Travel and Tourism.* Green River and fall foliage *courtesy of Superstock.*

VIRGINIA

Courtesy of Superstock.

WASHINGTON

Courtesy of Superstock.

WEST VIRGINIA

Courtesy of West Virginia Division of Tourism. Harper's Ferry, Endless Wall, Blackwater Falls, Pillow Rock Rapid, and Beckley Coal Mine, *Stephen Shaluta, Jr., photographer;* Seneca Rocks, *David Fattaléh, photographer.*

WISCONSIN

Cows, capitol building, cross-country skiing, and Hawk's Beak *courtesy of Wisconsin Division of Tourism.* Scenic view *courtesy of Superstock.*

WYOMING

Shoshone Boys, trail ride, Old Faithful, Old Town Trail, and Cheyenne Frontier Days *courtesy of the Wyoming Division of Tourism.* Devil's Tower National Monument *courtesy of Superstock.*

WASHINGTON, D.C.

Courtesy of Superstock.

U.S. POSSESSIONS:

American Samoa coastline *courtesy of Office of Tourism—American Samoa;* Cruz Bay *courtesy of U.S. Virgin Islands Division of Tourism.* El Morro Fortress and San Juan Jeronimo Fort *courtesy of Superstock.*

ALBERTA

Courtesy of Photo Search Ltd., John Sutton, photographer.

BRITISH COLUMBIA

Courtesy of Superstock.

MANITOBA

Polar bear *courtesy of Ed Struzik, Northwest Territories Economic Development and Tourism.* Northern Boreal Forest, Fort Garry, Brandon, and the Legislative Building *courtesy of Superstock.*

NEW BRUNSWICK

Miramichi River, covered bridge, kayaking, St. John, and King's Landing *courtesy of New Brunswick Tourism.* Point-du-Chene *courtesy of Superstock.*

NEWFOUNDLAND

Courtesy of the Department of Tourism and Culture, Newfoundland and Labrador.

NORTHWEST TERRITORIES

Courtesy of Northwest Territories Economic Development and Tourism. Inuk girl and Yellowknife, *Doug Walker, photographer;* iceberg, *Dan Heringa, photographer;* dog team, *Wolfgang Weber, photographer.*

NOVA SCOTIA

Whale watching and rock formations *courtesy of Nova Scotia Department of Information and Communications.* Neils Harbor, Peggy's Cove, Halifax, and Cabot Trail *courtesy of Superstock.*

ONTARIO

Toronto City Hall, the Sky Dome, and Toronto skyline *courtesy of the Metropolitan Toronto Convention and Visitors Association.* Niagara Falls and the Parliament Buildings *courtesy of Superstock.*

PRINCE EDWARD ISLAND

Courtesy of Superstock.

QUÉBEC

Courtesy of Superstock.

SASKATCHEWAN

Courtesy of Tourism Saskatchewan.

YUKON TERRITORY

Courtesy of Superstock.

Map Contents

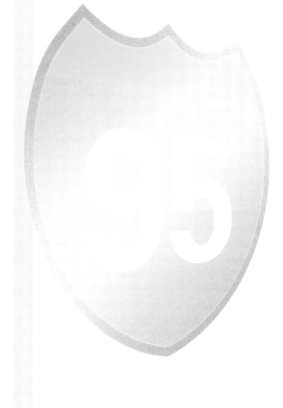

Legend

Roads

Freeway	(95) Interstate route
Tollway	(60) US route
Under construction	(73) State or provincial route
Divided highway	
Primary road	(12) Other route
Secondary road	(2) Mexican Federal route
Other road	(00) Canadian auto route
Unpaved road	
150 Interchange & exit #	Trans-Canada route

Boundaries

State boundary
International boundary

Cities with population...

Phoenix⊙ over 500,000
Glendale⊙ 100,000 - 499,999
Sun City⊙ 25,000 - 99,999
Buckeye○ 5,000 - 24,999
Avondale○ 0 - 4,999

Parks etc.

National park
National forest or other recreational area
Military lands
Indian reservation

Symbols

+ Mountain peak
≍ Pass
▪ Point of interest

Built up area
✱ State Capital
✪ National Capital

◄ 10 ► Distance in miles along primary route

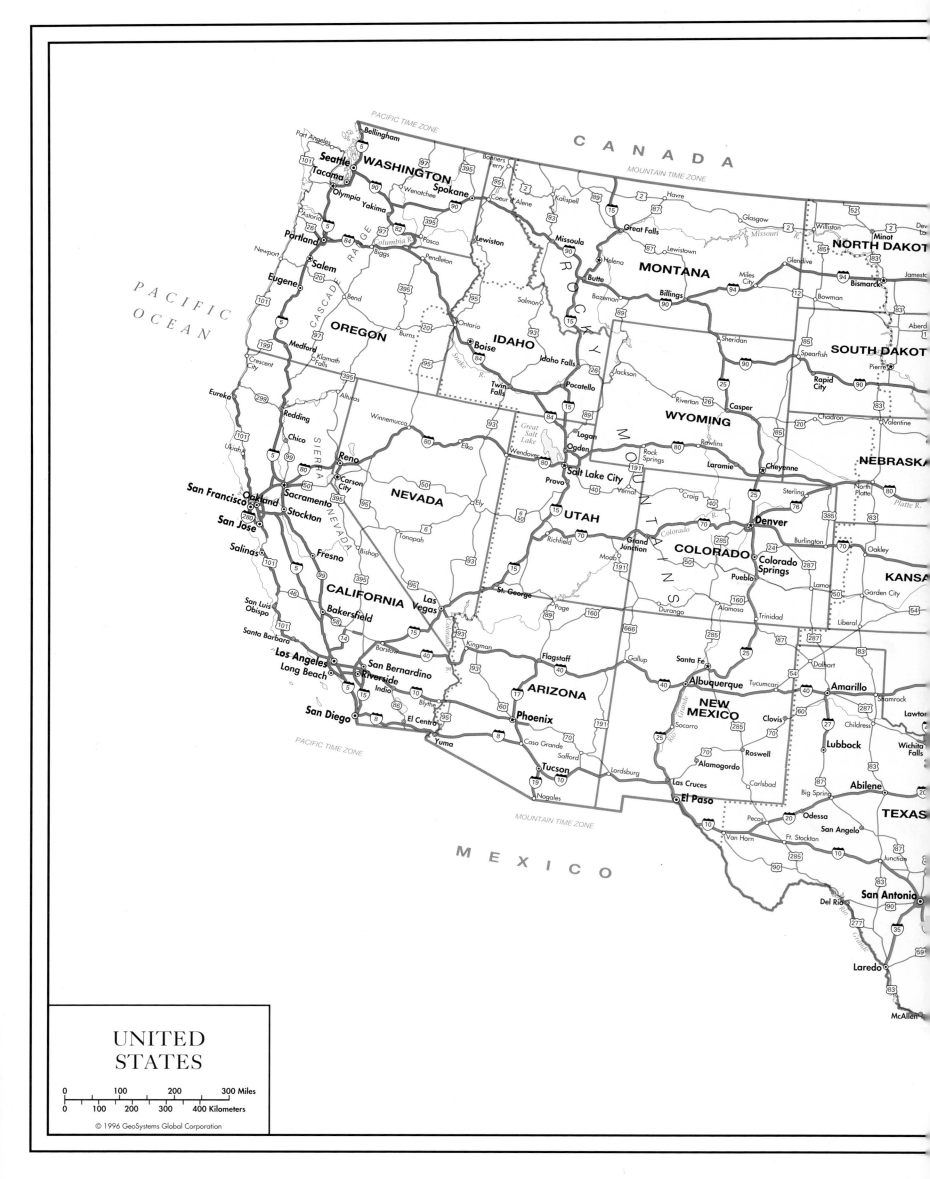

UNITED STATES

```
0        100       200       300 Miles
0    100    200    300    400 Kilometers
```

© 1996 GeoSystems Global Corporation

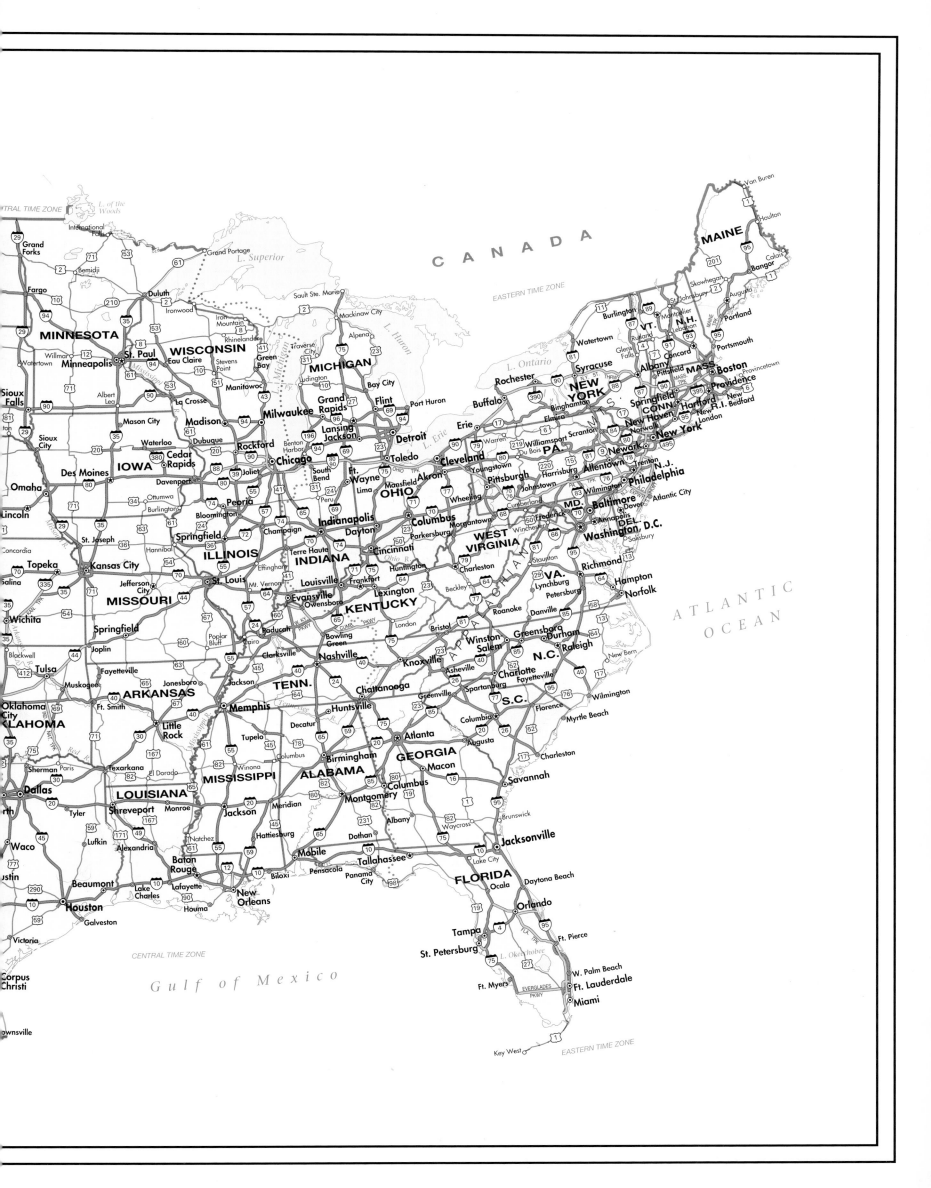

UNITED STATES

	Albuquerque, N. MEX.	Atlanta, GA.	Bangor, MAINE	Billings, MONT.	Boston, MASS.	Charlotte, N.C.	Chicago, ILL.	Cleveland, OHIO	Dallas, TEX.	Denver, COLO.	Detroit, MICH.	Kansas City, MO.	Los Angeles, CALIF.
Albuquerque, N. MEX.		1409	2490	992	2236	1665	1353	1577	665	444	1547	795	80
Atlanta, GA.	1409		1331	1831	1044	238	717	725	791	1403	730	800	225
Bangor, MAINE	2490	1331		2498	238	1093	1243	883	2050	2244	942	1676	327
Billings, MONT.	992	1831	2498		2254	2025	1264	1601	1353	555	1532	1077	123
Boston, MASS.	2236	1044	238	2254		864	990	659	1781	1985	722	1441	304
Charlotte, N.C.	1665	238	1093	2025	864		772	520	1031	1575	675	972	247
Chicago, ILL.	1353	717	1243	1264	990	772		346	936	1013	282	530	204
Cleveland, OHIO	1577	725	883	1601	659	520	346		1208	1347	172	806	237
Dallas, TEX.	665	791	2050	1353	1781	1031	936	1208		797	1218	518	144
Denver, COLO.	444	1403	2244	555	1985	1575	1013	1347	797		1283	606	103
Detroit, MICH.	1547	730	942	1532	722	675	282	172	1218	1283		752	231
Kansas City, MO.	795	800	1676	1077	1441	972	530	806	518	606	752		162
Los Angeles, CALIF.	806	2258	3273	1236	3044	2479	2041	2375	1446	1030	2311	1628	
Memphis, TENN.	1034	382	1584	1527	1330	630	539	742	466	1069	752	450	184
Miami, FLA.	1951	665	1756	2516	1527	729	1350	1248	1370	2081	1395	1462	274
Minneapolis, MINN.	1250	1130	1661	838	1432	1188	409	760	999	924	679	442	193
Nashville, TENN.	1238	242	1367	1613	1096	413	473	530	681	1161	541	558	205
New Orleans, LA.	1175	472	1793	1863	1529	713	935	1055	510	1307	1067	819	194
New York, N.Y.	2014	868	448	2051	204	630	796	466	1587	1798	622	1255	282
Orlando FLA.	1934	440	1552	2333	1323	524	1161	1044	1144	1847	1179	1244	253
Phoenix, ARIZ.	468	1825	2957	1202	2687	2109	1750	2028	1025	787	1998	1246	37
Portland, OREG.	1355	2708	3368	889	3139	2821	2118	2449	2142	1257	2386	1848	97
Rapid City, S.D.	817	1511	2166	358	1913	1693	913	1254	1080	404	1191	710	139
St. Louis, MO.	1050	575	1429	1340	1200	720	288	559	634	861	548	252	185
Salt Lake City, UTAH	607	1923	2643	556	2390	2025	1411	1745	1267	538	1668	1122	69
San Antonio, TEX.	818	999	2323	1503	2094	1240	1226	1481	271	960	1491	795	135
San Francisco, CALIF.	1109	2588	3375	1241	3141	2774	2151	2482	1830	1255	2399	1861	38
Seattle, WASH.	1454	2653	3309	813	3056	2839	2060	2414	2119	1329	2351	1902	115
Tampa, FLA.	1951	456	1608	2350	1379	581	1177	1101	1161	1863	1196	1261	255
Washington, D.C.	1871	635	688	1963	460	397	703	372	1307	1679	525	1077	269

MILEAGE CHART

	Memphis, TENN.	Miami, FLA.	Minneapolis, MINN.	Nashville, TENN.	New Orleans, LA.	New York, N.Y.	Orlando FLA.	Phoenix, ARIZ.	Portland, OREG.	Rapid City, S.D.	St. Louis, MO.	Salt Lake City, UTAH	San Antonio, TEX.	San Francisco, CALIF.	Seattle, WASH.	Tampa, FLA.	Washington, D.C.
034	1951	1250	1238	1175	2014	1934	468	1355	817	1050	607	818	1109	1454	1951	1871
382	665	1130	242	472	868	440	1825	2708	1511	575	1923	999	2588	2653	456	635
584	1756	1661	1367	1793	448	1552	2957	3368	2166	1429	2643	2323	3375	3309	1608	688
527	2516	838	1613	1863	2051	2333	1202	889	358	1340	556	1503	1241	813	2350	1963
330	1527	1432	1096	1529	204	1323	2687	3139	1913	1200	2390	2094	3141	3056	1379	460
530	729	1188	413	713	630	524	2109	2821	1693	720	2025	1240	2774	2839	581	397
539	1350	409	473	935	796	1161	1750	2118	913	288	1411	1226	2151	2060	1177	703
742	1248	760	530	1055	466	1044	2028	2449	1254	559	1745	1481	2482	2414	1101	372
466	1370	999	681	510	1587	1144	1025	2142	1080	634	1267	271	1830	2119	1161	1307
069	2081	924	1161	1307	1798	1847	787	1257	404	861	538	960	1255	1329	1863	1679
752	1395	679	541	1067	622	1179	1998	2386	1191	548	1668	1491	2399	2351	1196	525
450	1462	442	558	819	1255	1244	1246	1848	710	252	1122	795	1861	1902	1261	1077
840	2744	1932	2055	1943	2820	2535	374	971	1394	1857	691	1356	387	1150	2552	2697
	1007	941	215	397	1092	900	1476	2313	1169	293	1573	739	2143	2419	917	856
007		1761	903	876	1342	233	2391	3306	2181	1218	2593	1403	3130	3363	289	1060
941	1761		858	1337	1211	1577	1684	1727	580	621	1322	1257	2034	1655	1594	1118
215	903	858		539	904	685	1676	2407	1255	337	1681	954	2358	2461	702	656
397	876	1337	539		1330	650	1525	2530	1560	689	1777	557	2301	2665	667	1089
092	1342	1211	904	1330		1089	2482	2921	1719	982	2196	1860	2927	2901	1145	236
900	233	1577	685	650	1089		2165	3092	1955	992	2367	1178	2919	3146	81	855
476	2391	1684	1676	1525	2482	2165		1293	1160	1501	650	986	757	1520	2182	2322
313	3306	1727	2407	2530	2921	3092	1293		1268	2100	753	2098	636	173	3109	2861
69	2181	580	1255	1560	1719	1955	1160	1268		962	746	1230	1454	1160	1971	1605
293	1218	621	337	689	982	992	1501	2100	962		1375	916	2110	2154	1009	831
573	2593	1322	1681	1777	2196	2367	650	753	746	1375		1425	749	852	2384	2108
739	1403	1257	954	557	1860	1178	986	2098	1230	916	1425		1740	2278	1194	1635
43	3130	2034	2358	2301	2927	2919	757	636	1454	2110	749	1740		811	2936	2859
419	3363	1655	2461	2665	2901	3146	1520	173	1160	2154	852	2278	811		3163	2765
917	289	1594	702	667	1145	81	2182	3109	1971	1009	2384	1194	2936	3163		912
56	1060	1118	656	1089	236	855	2322	2861	1605	831	2108	1635	2859	2765	912	

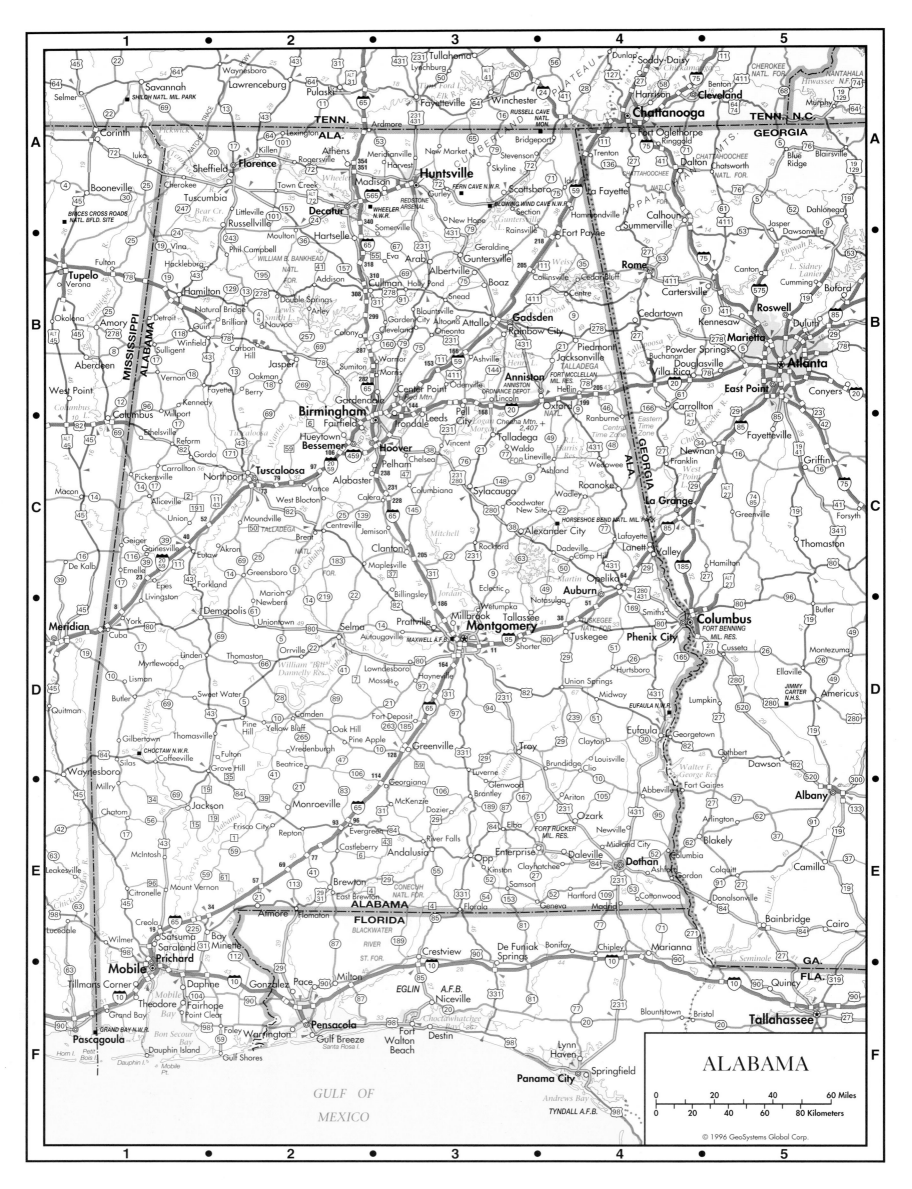

ALABAMA

Scale: 0 20 40 60 Miles
0 20 40 60 80 Kilometers

© 1996 GeoSystems Global Corp.

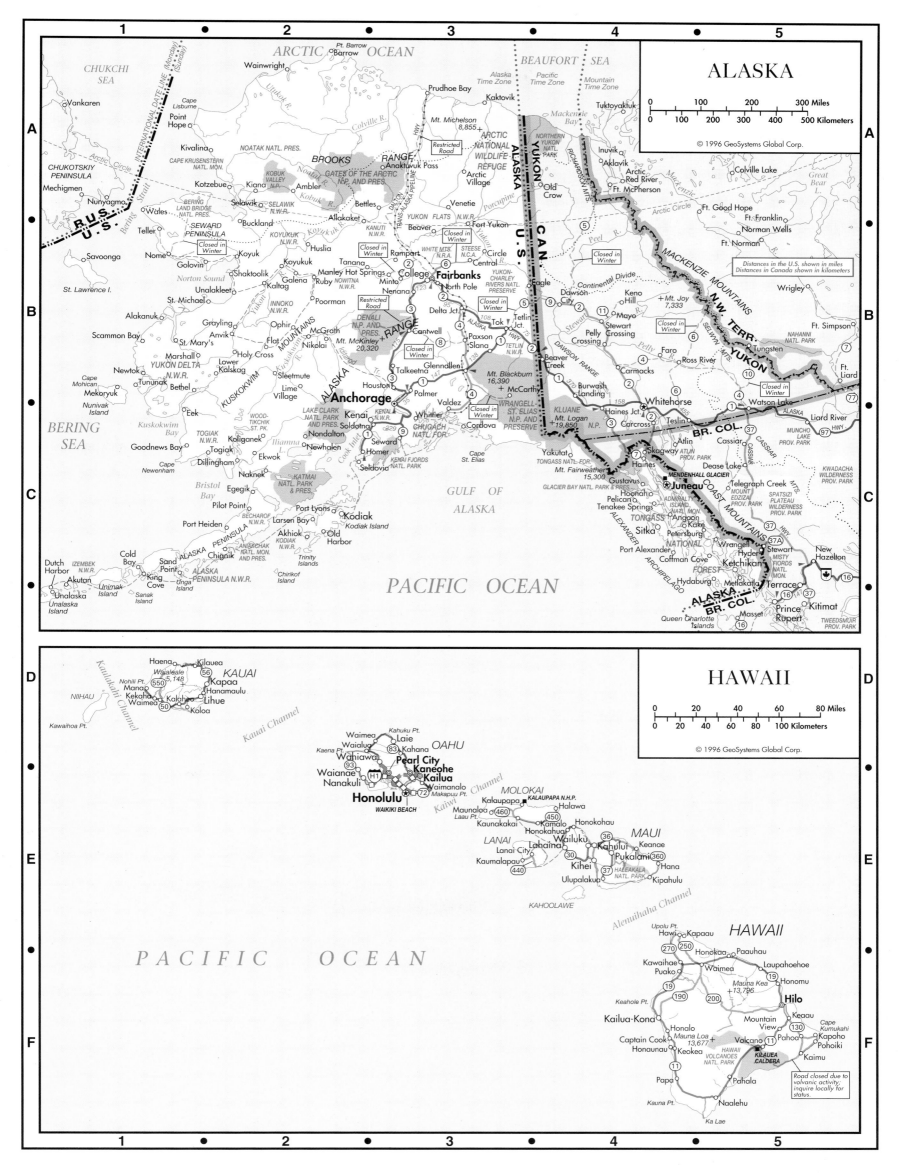

Alaska/Hawaii 103

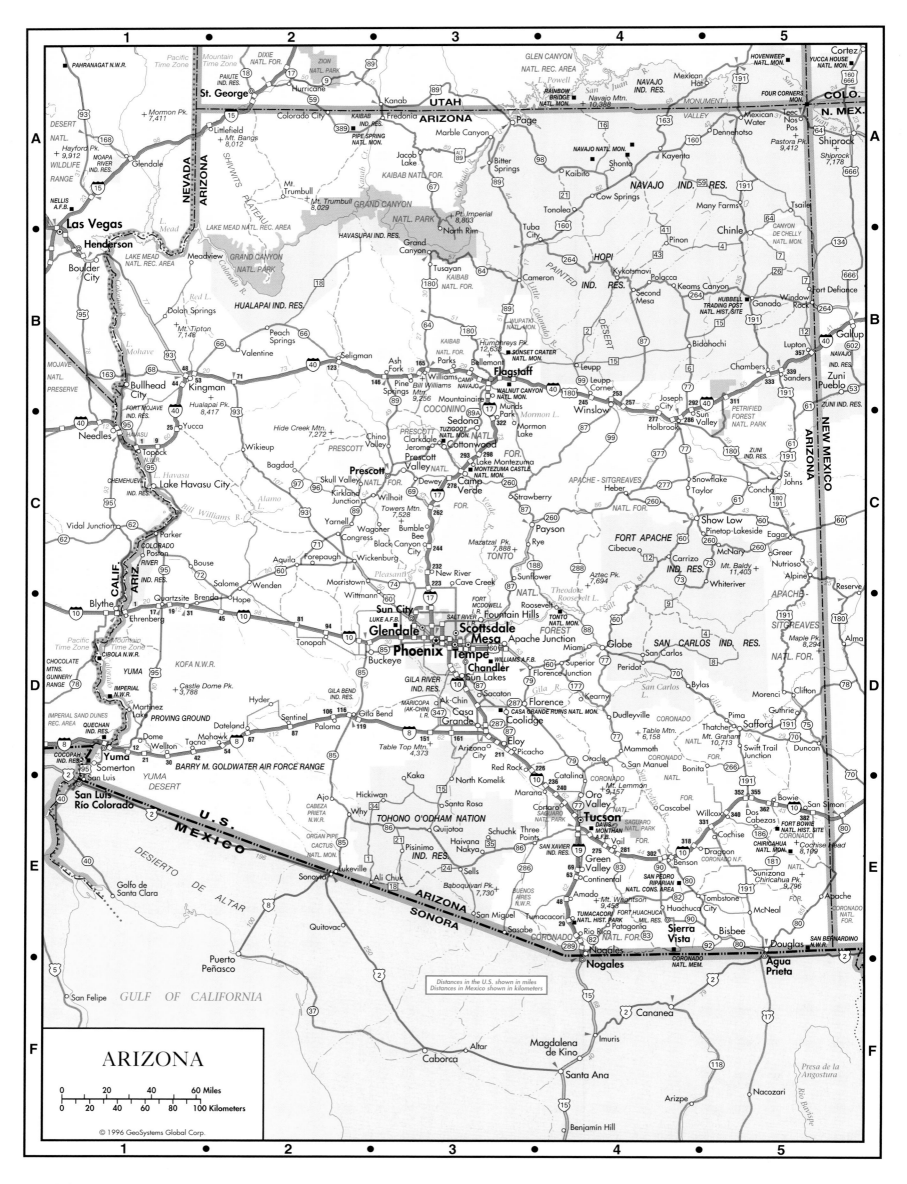

ARIZONA

0   20   40   60 Miles
0 20 40 60 80 100 Kilometers

© 1996 GeoSystems Global Corp.

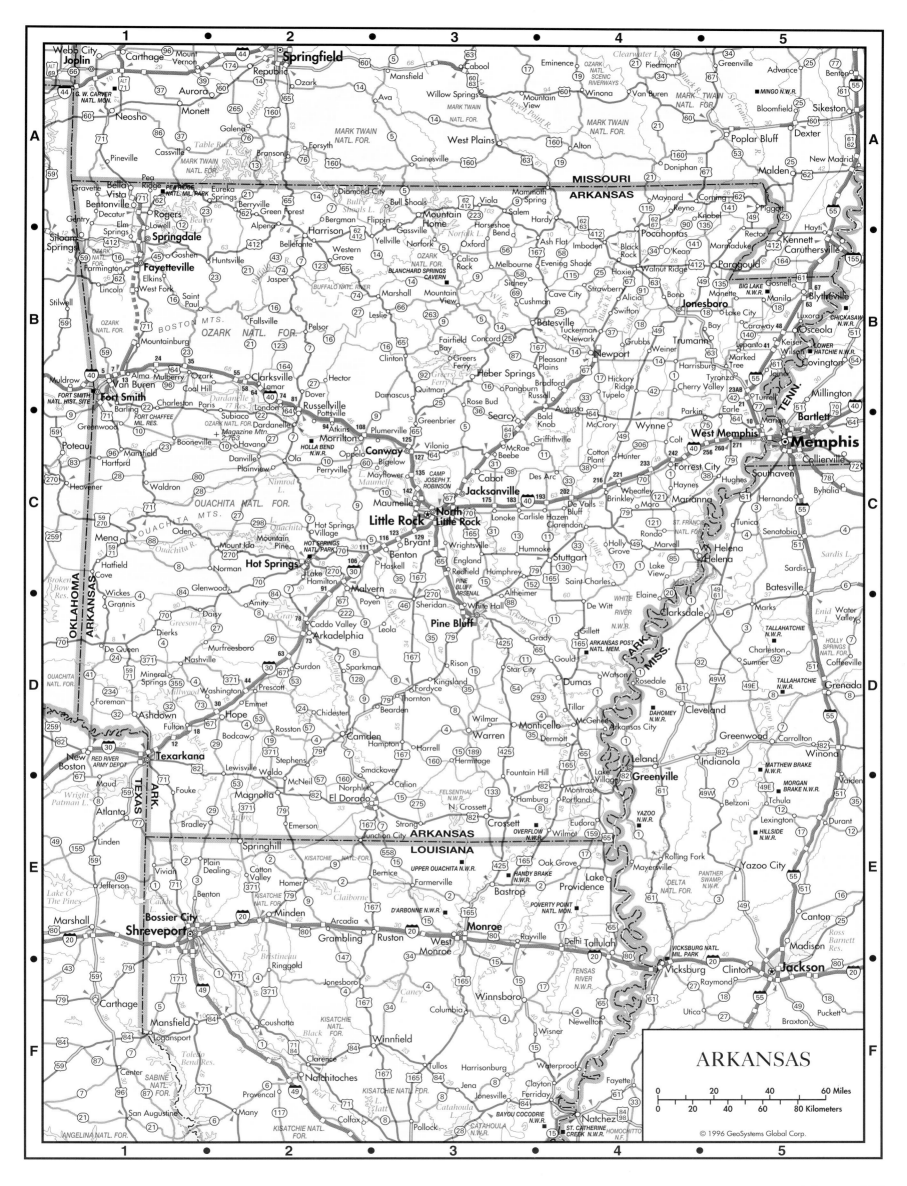

ARKANSAS

© 1996 GeoSystems Global Corp.

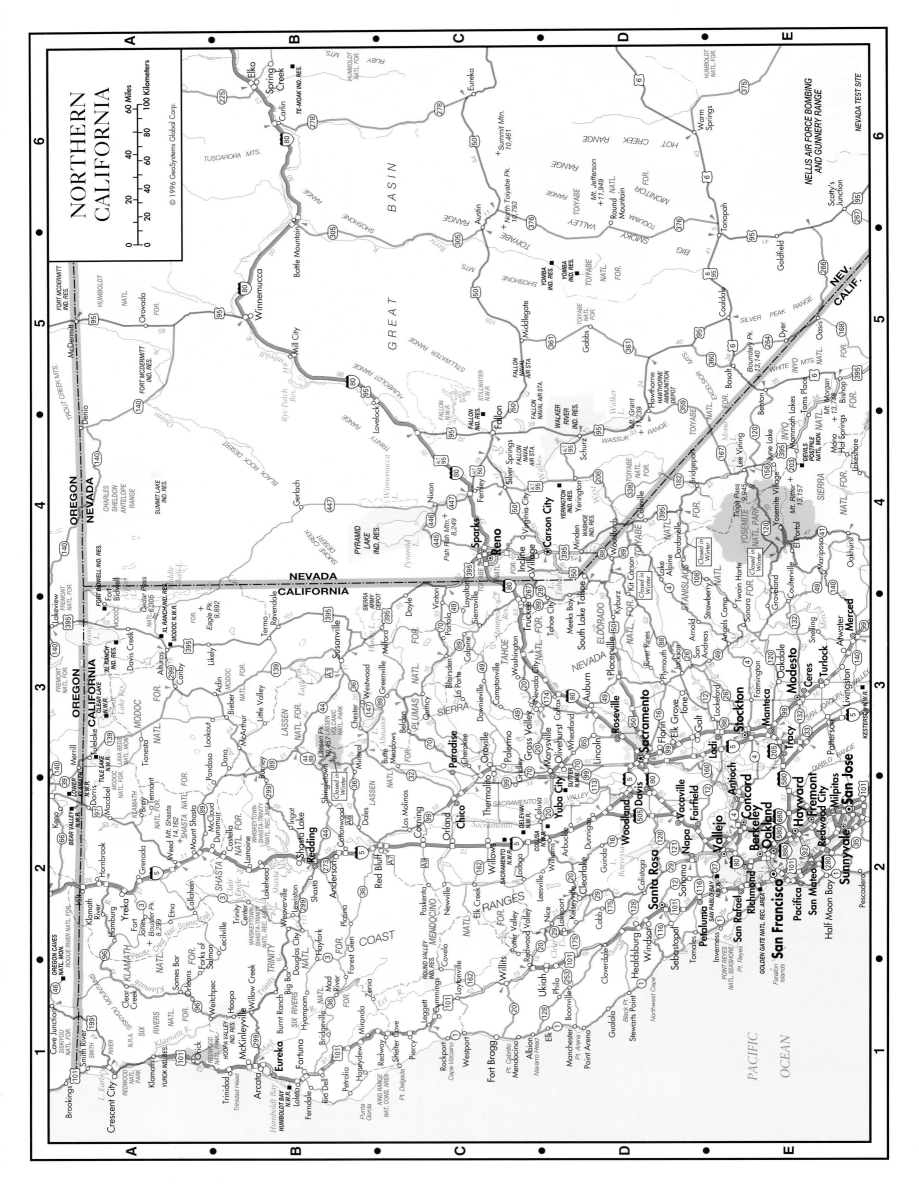

NORTHERN CALIFORNIA

© 1996 GeoSystems Global Corp.

60 Miles
100 Kilometers

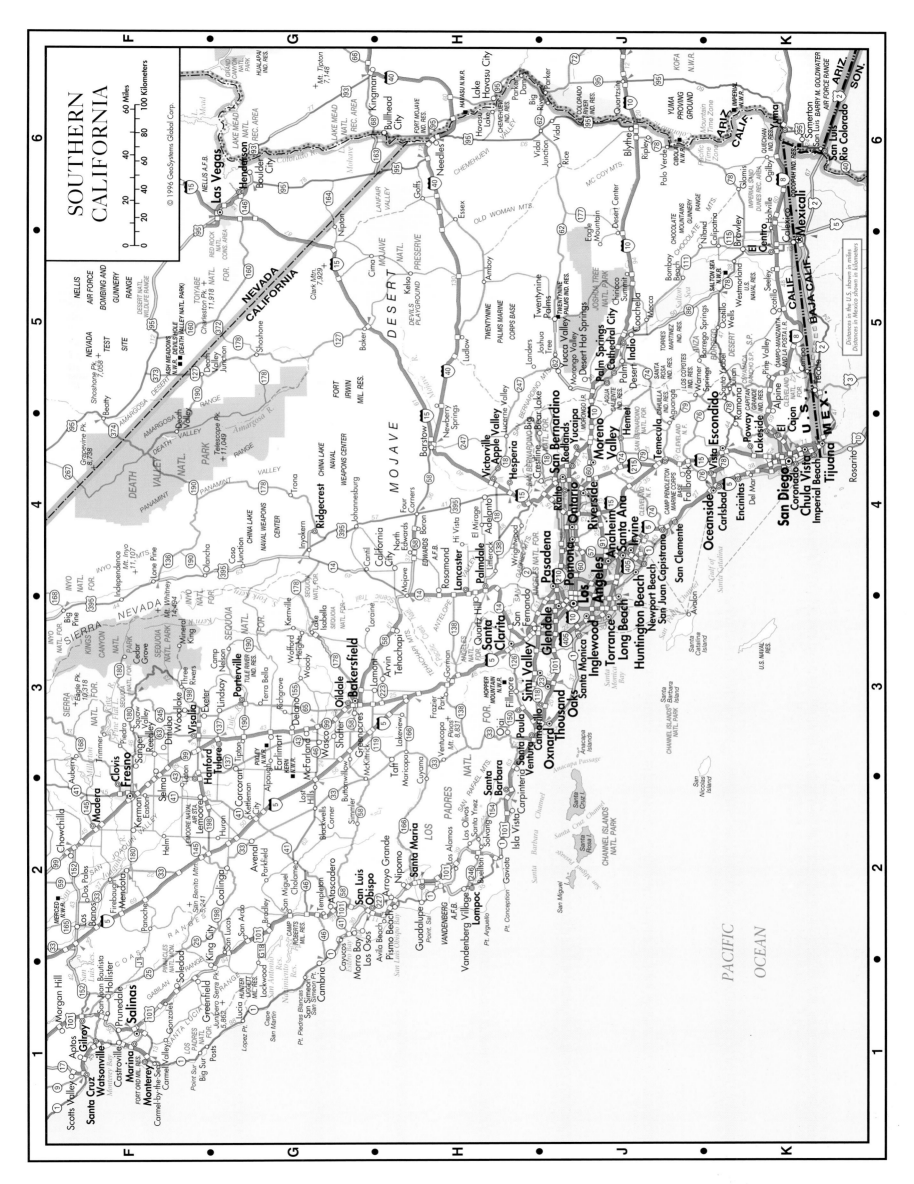

SOUTHERN CALIFORNIA

© 1996 GeoSystems Global Corp.

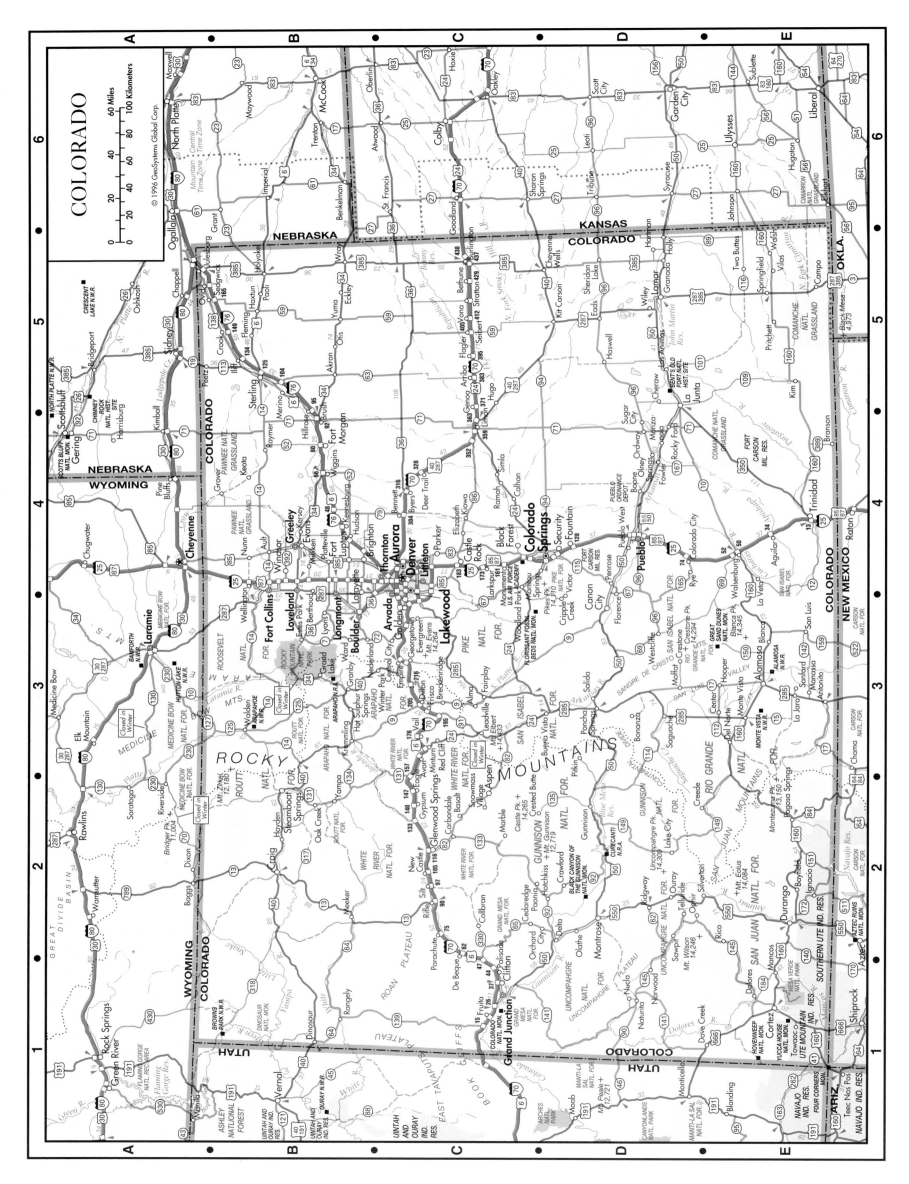

COLORADO

© 1996 GeoSystems Global Corp.

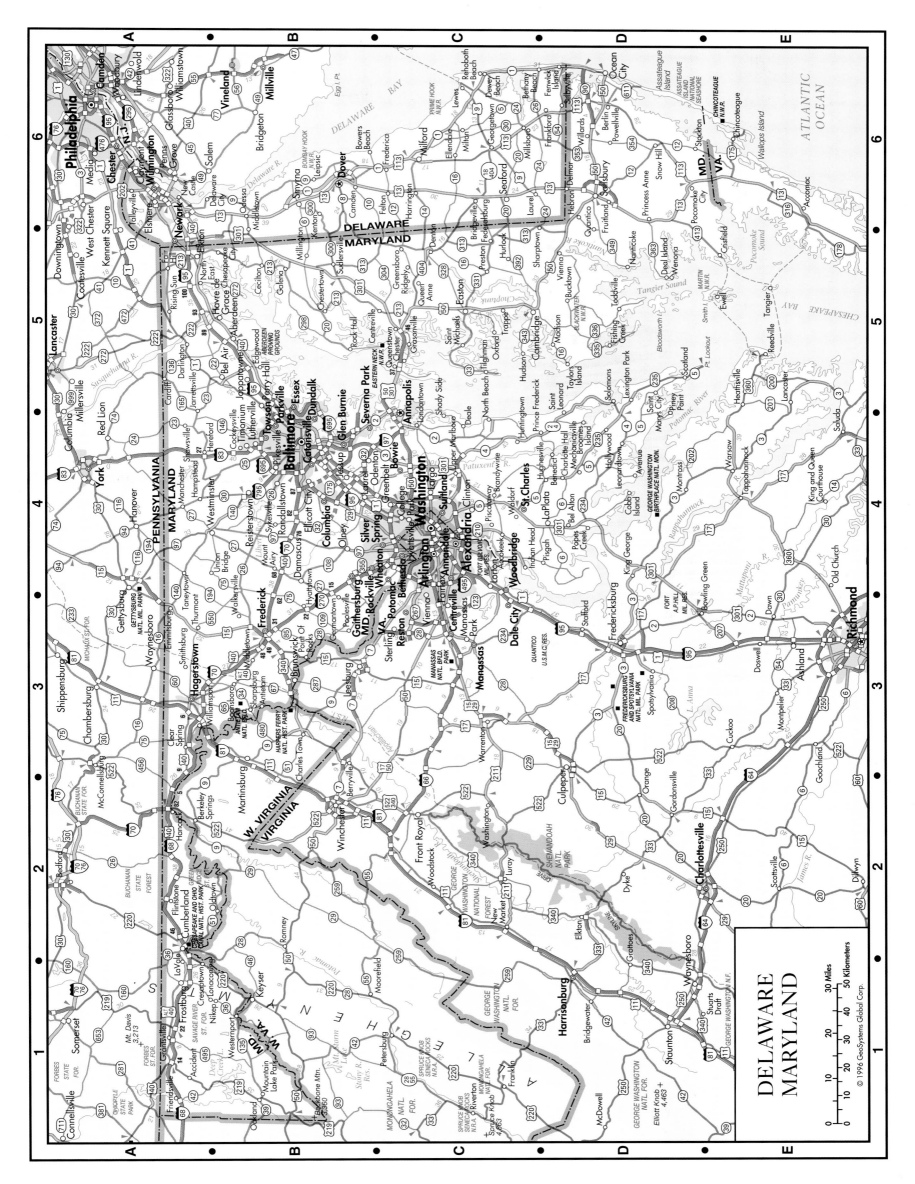

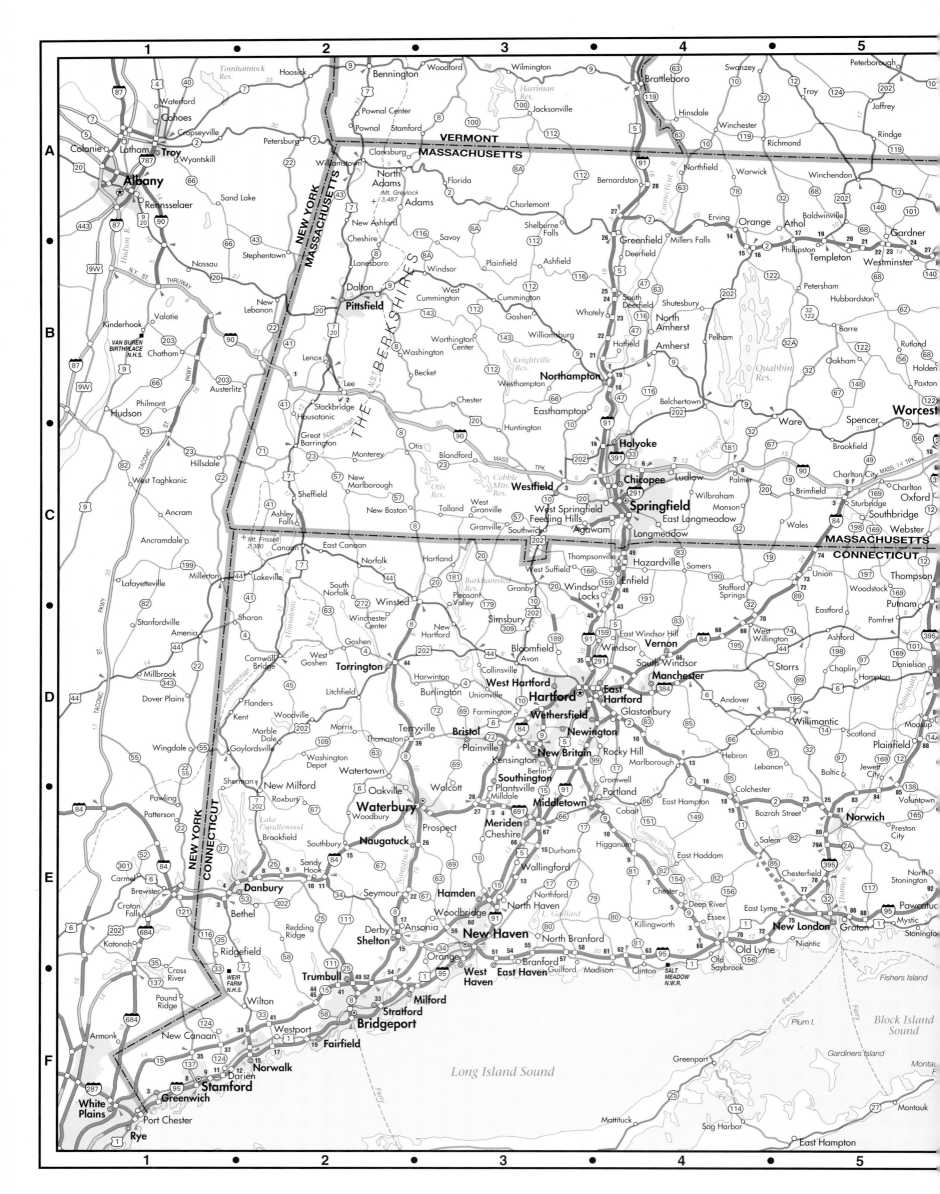

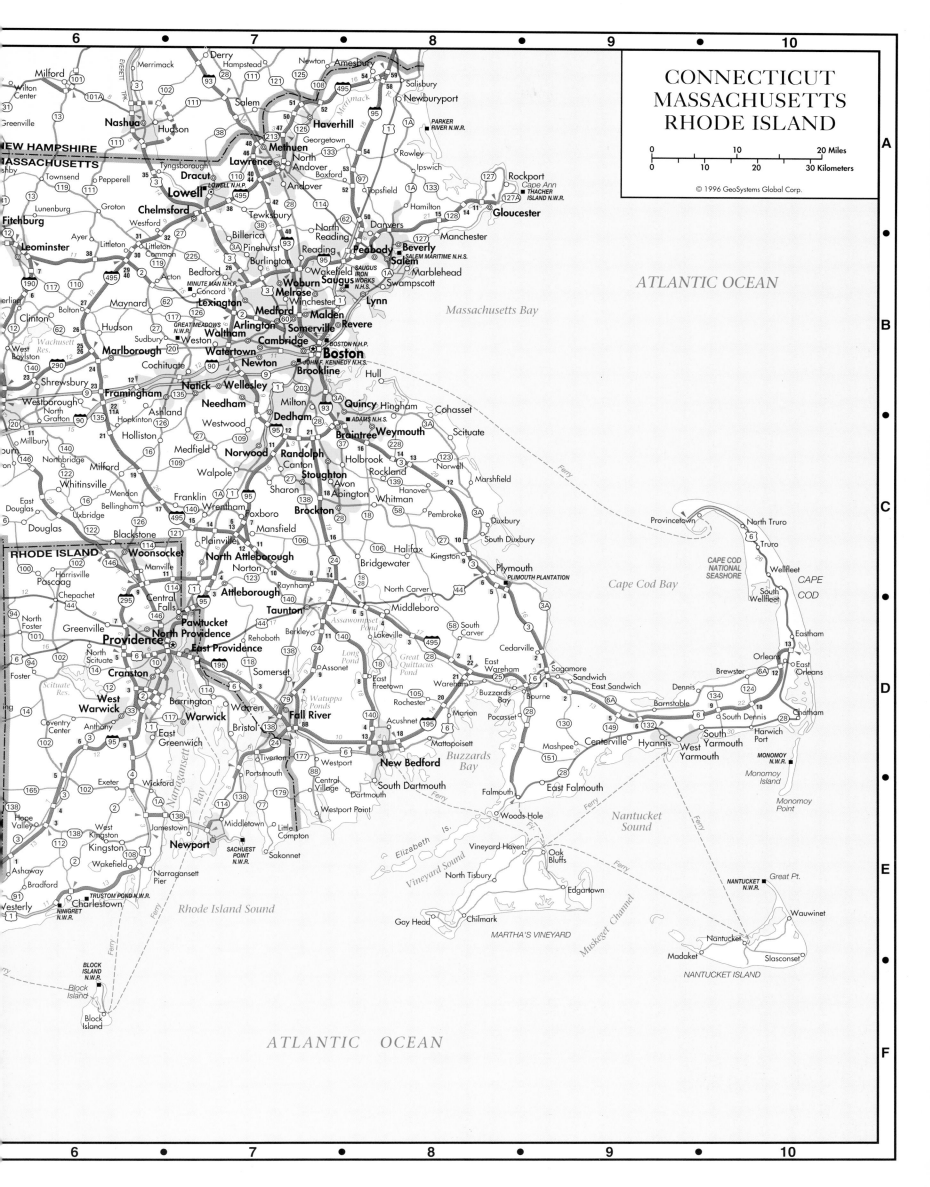

© 1996 GeoSystems Global Corp.

ATLANTIC OCEAN

Massachusetts Bay

Cape Cod Bay

CAPE COD NATIONAL SEASHORE

CAPE COD

Buzzards Bay

Nantucket Sound

Vineyard Sound

Muskeget Channel

MARTHA'S VINEYARD

NANTUCKET ISLAND

Rhode Island Sound

ATLANTIC OCEAN

NEW HAMPSHIRE
MASSACHUSETTS

RHODE ISLAND

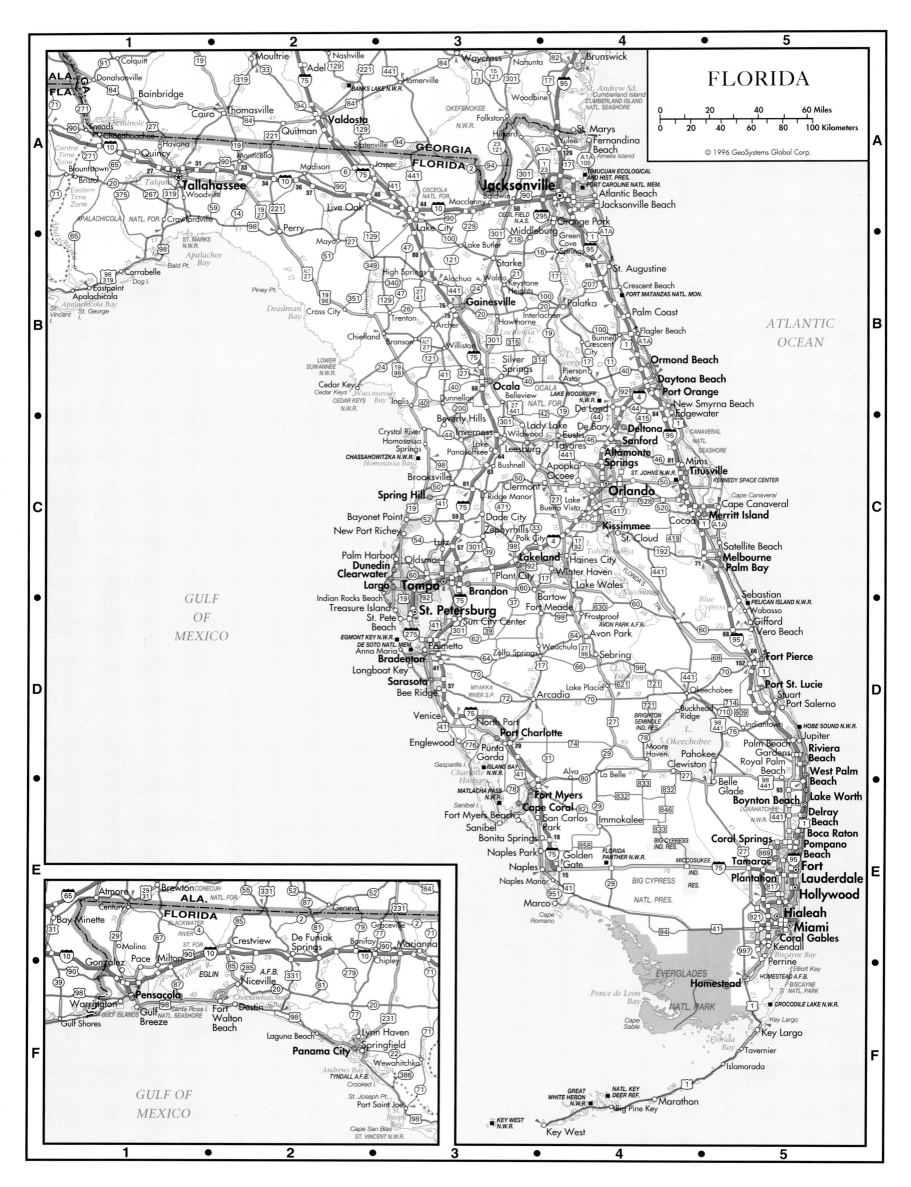

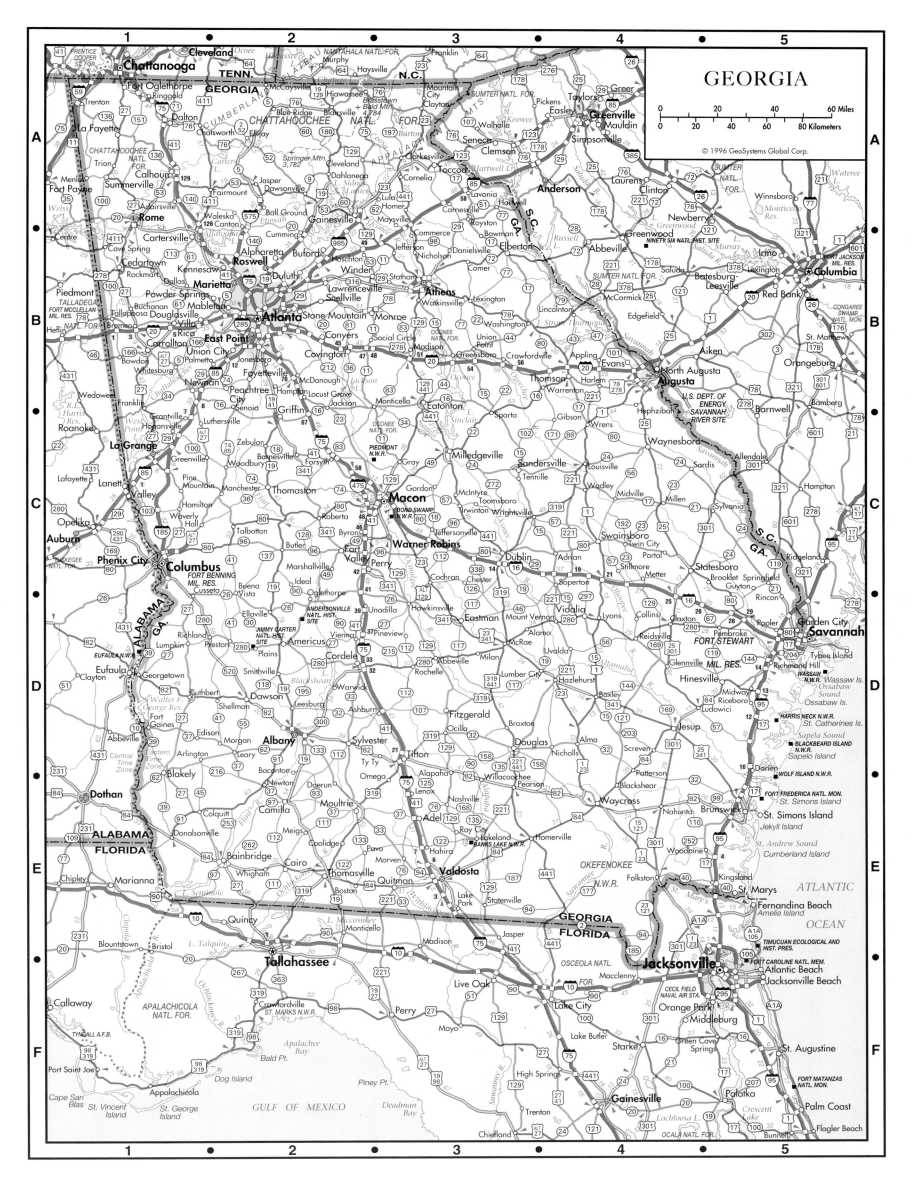

GEORGIA

0   20   40   60 Miles
0 20 40 60 80 Kilometers

© 1996 GeoSystems Global Corp.

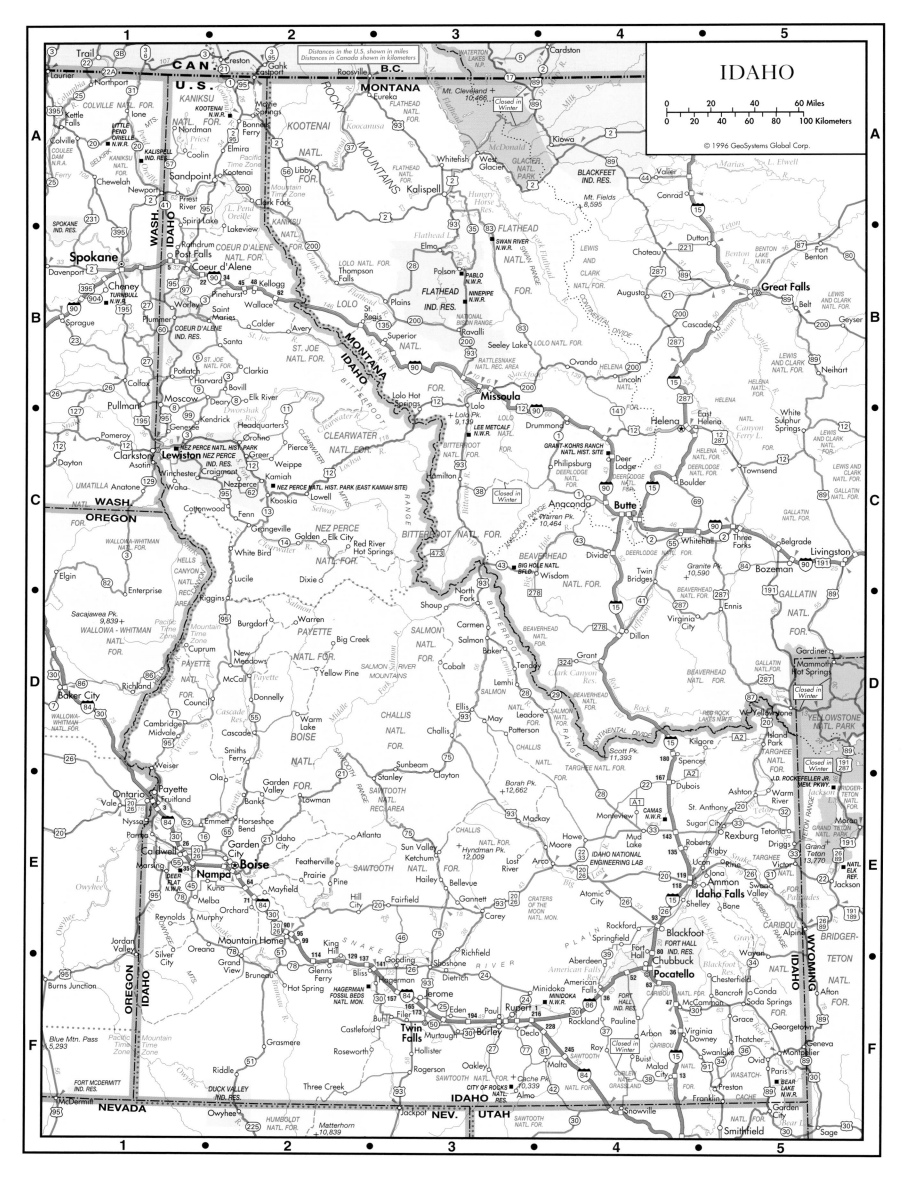

IDAHO

Distances in the U.S. shown in miles
Distances in Canada shown in kilometers

© 1996 GeoSystems Global Corp.

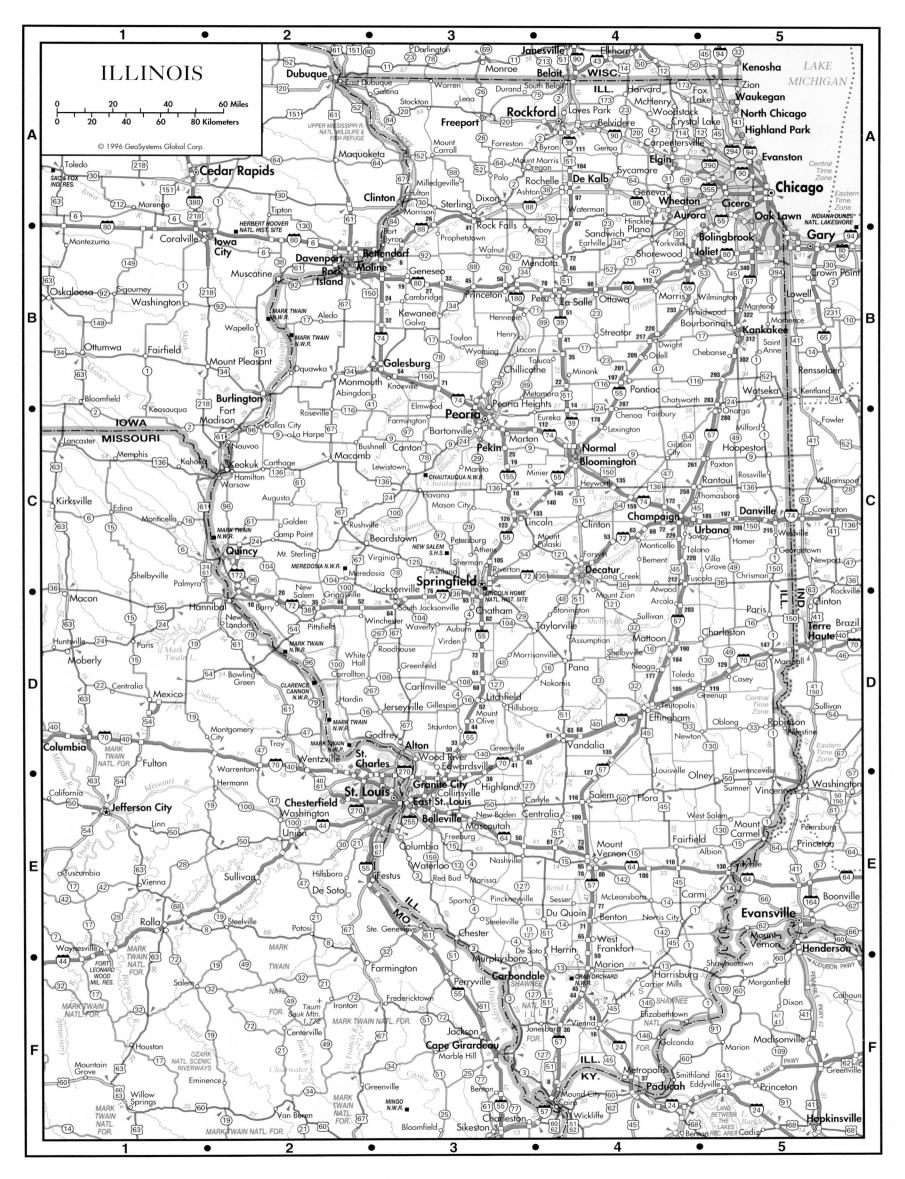

ILLINOIS

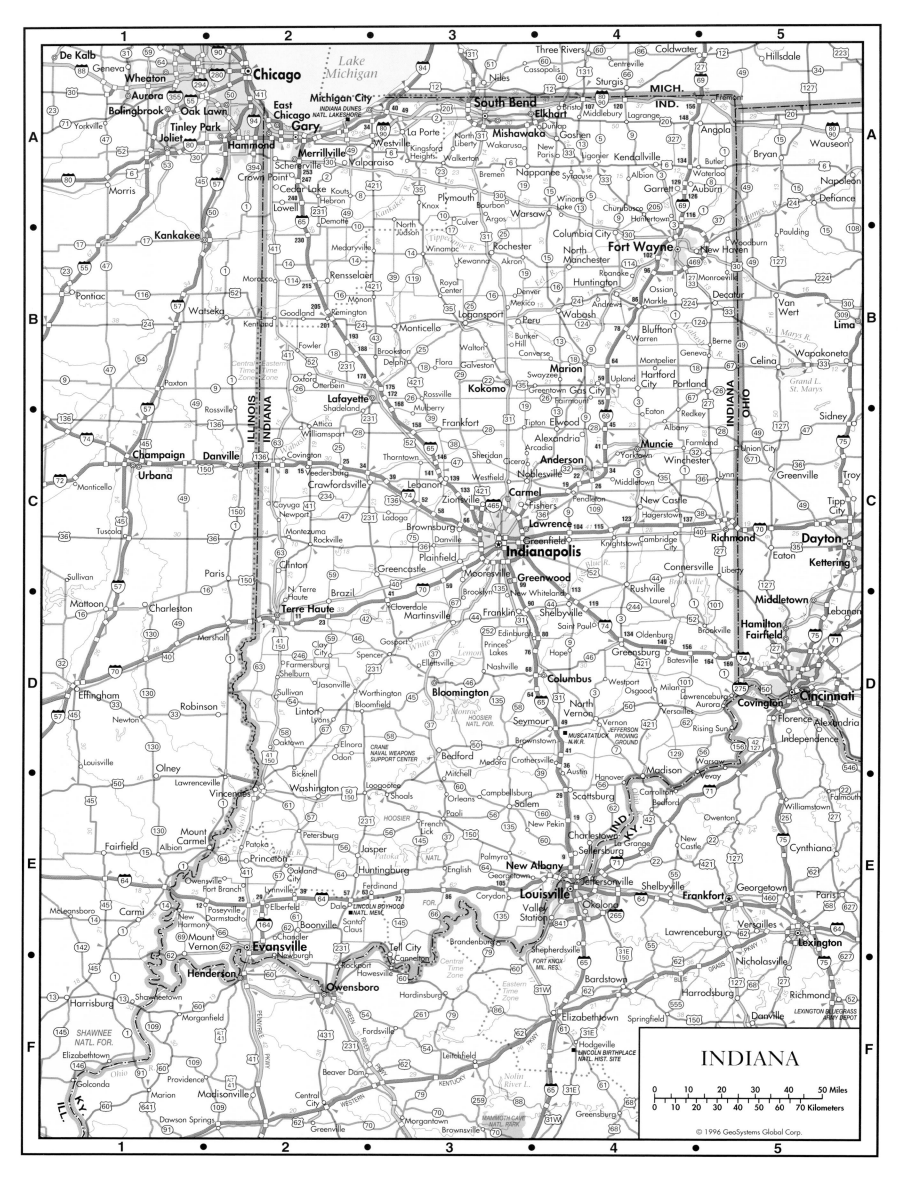

INDIANA

0 10 20 30 40 50 Miles
0 10 20 30 40 50 60 70 Kilometers

©1996 GeoSystems Global Corp.

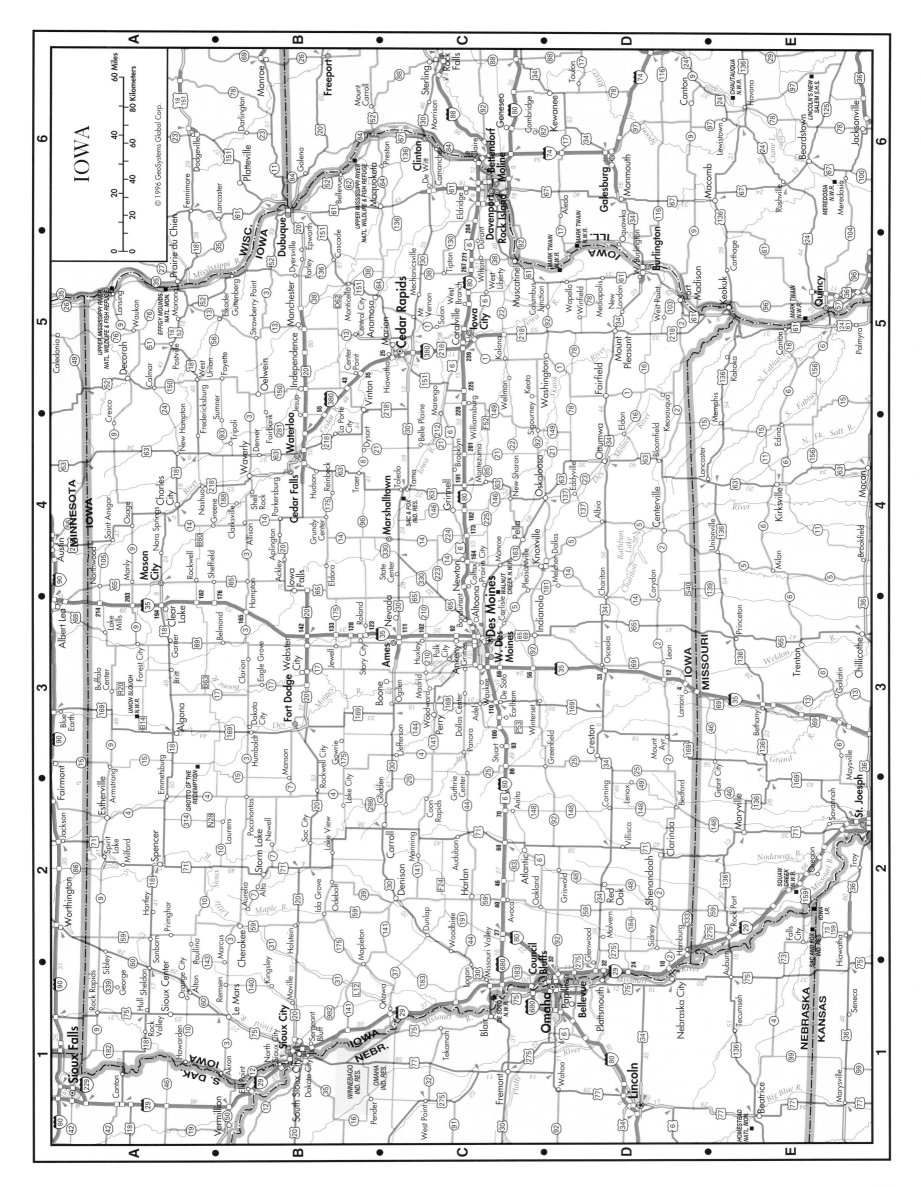

IOWA

© 1996 GeoSystems Global Corp.

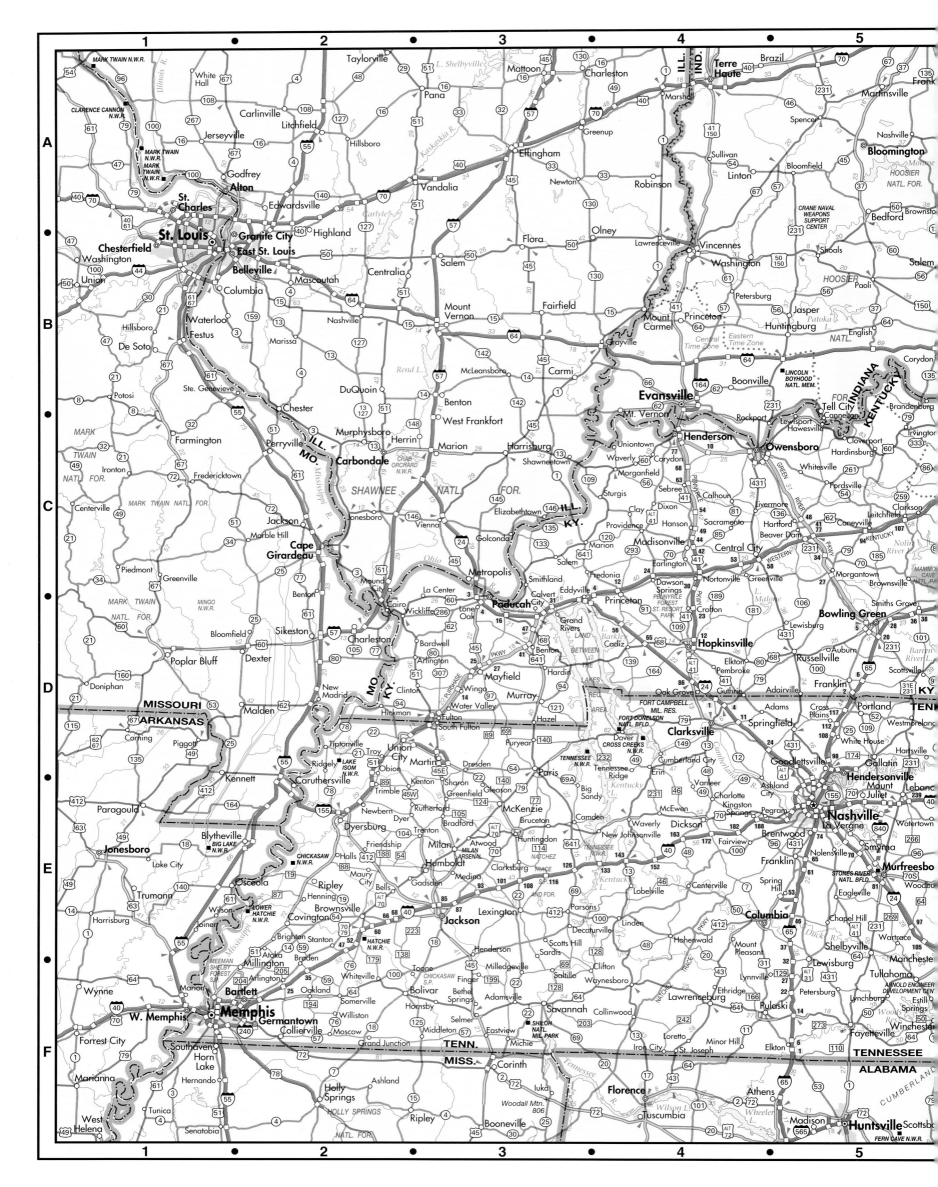

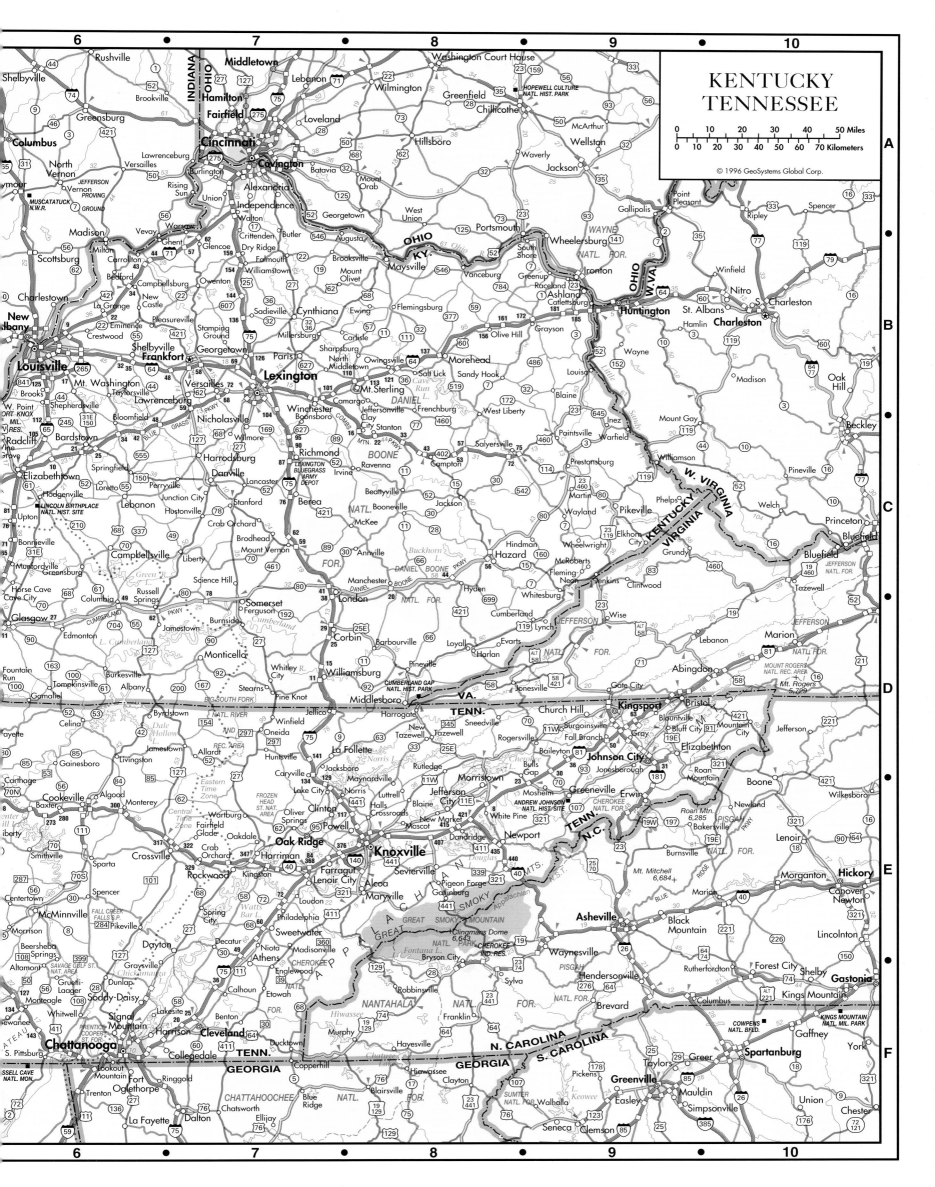

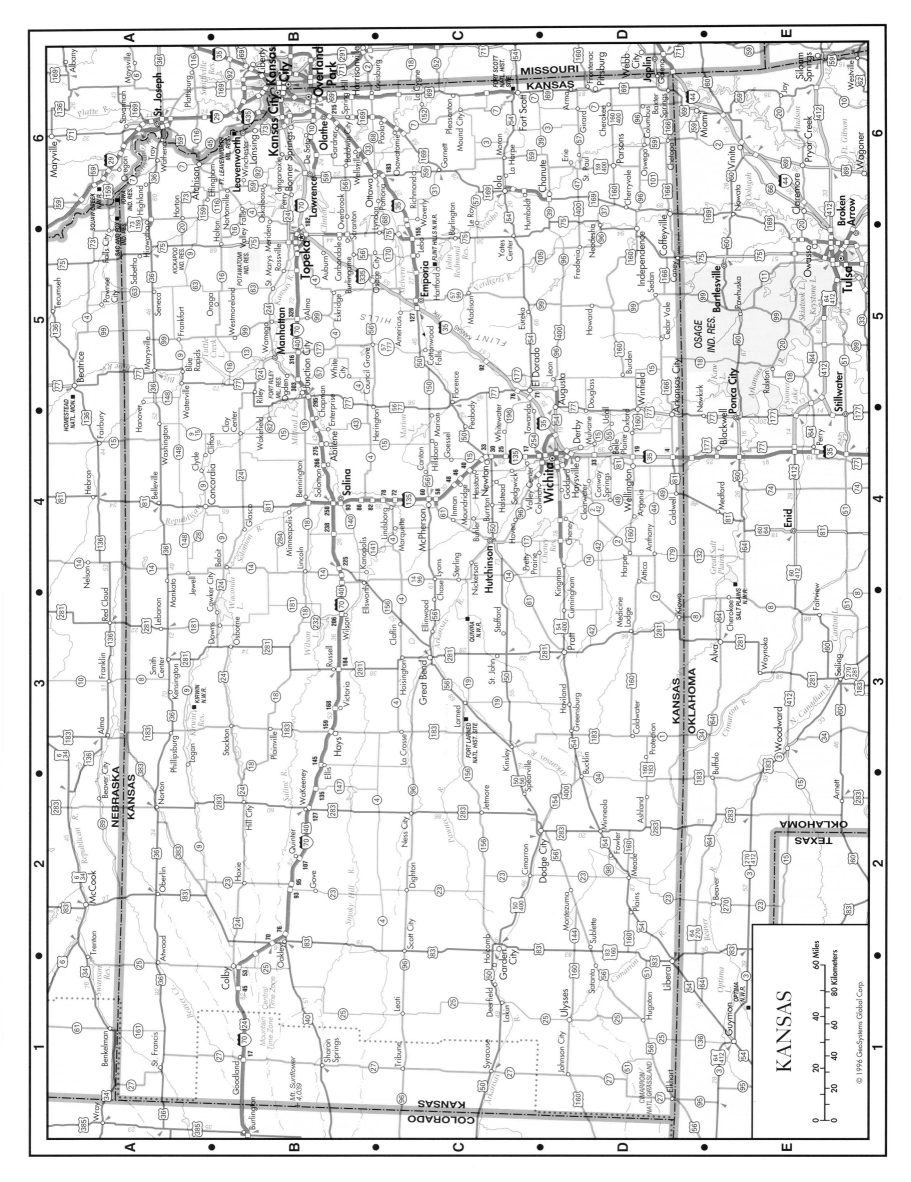

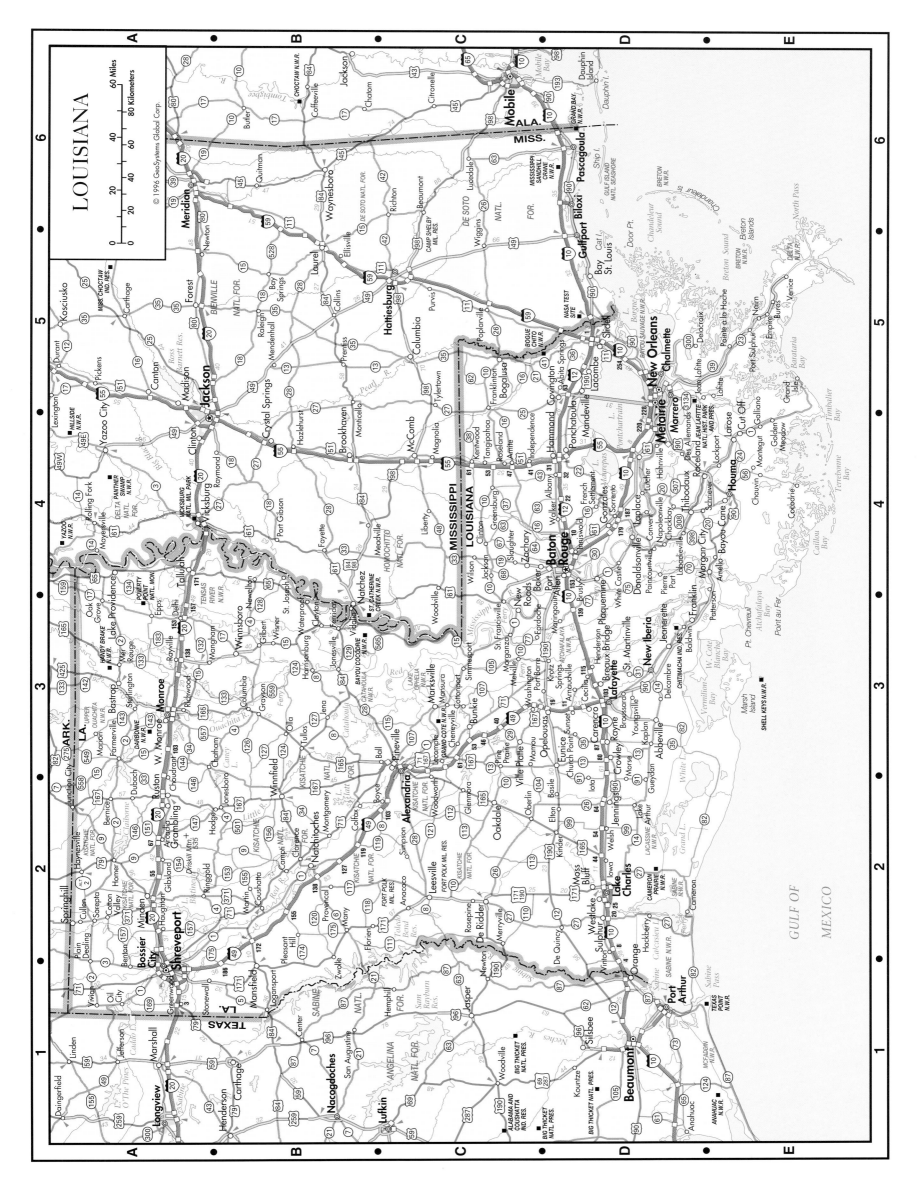

LOUISIANA

Louisiana 121

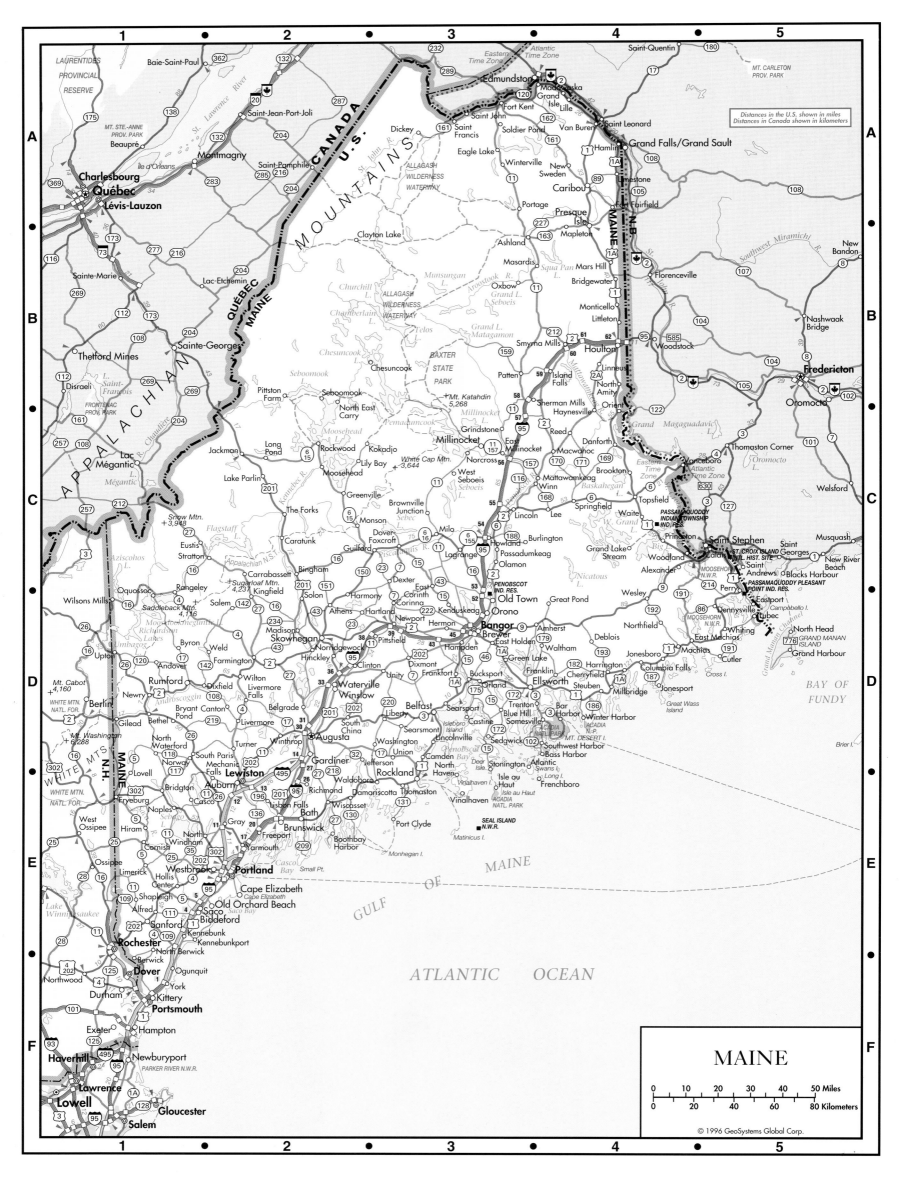

MAINE

Distances in the U.S. shown in miles
Distances in Canada shown in kilometers

0 10 20 30 40 50 Miles
0 20 40 60 80 Kilometers

© 1996 GeoSystems Global Corp.

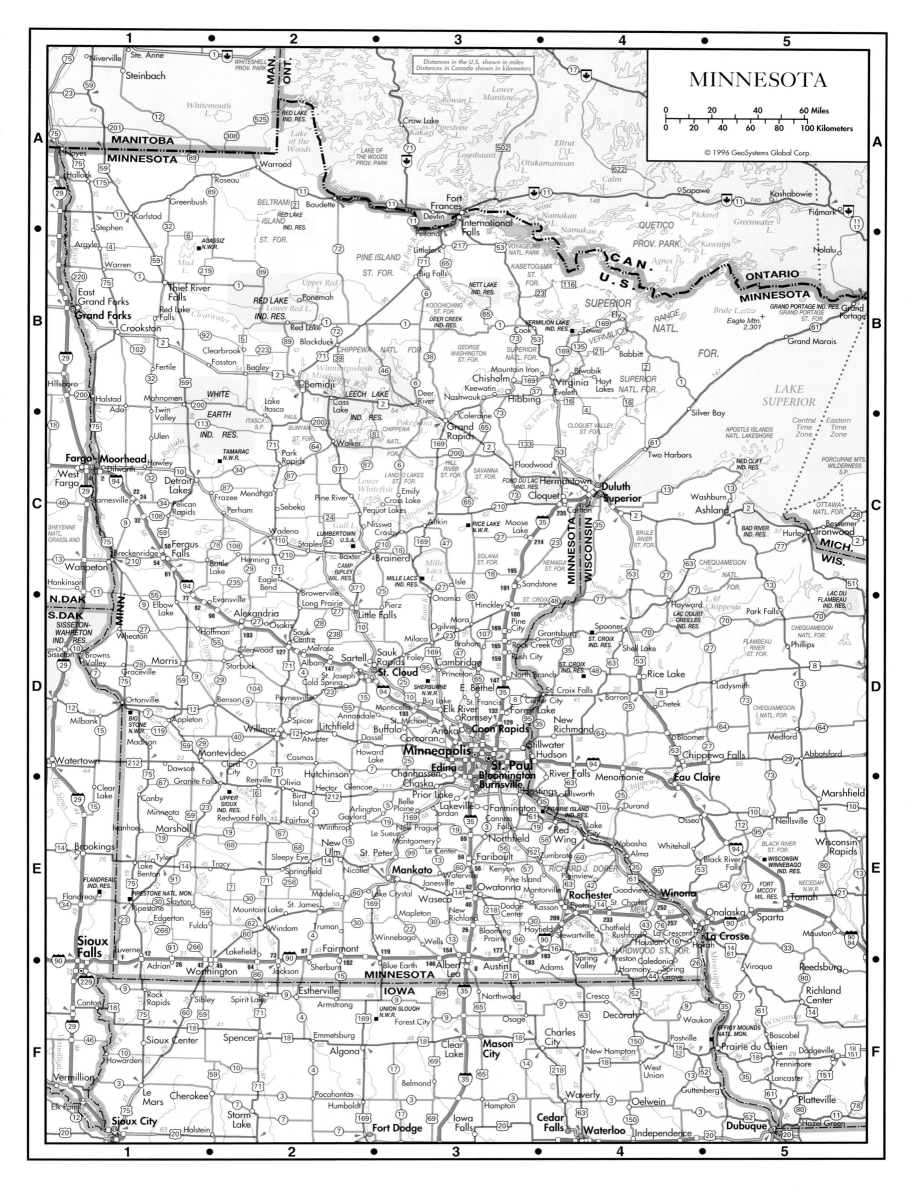

MINNESOTA

Distances in the U.S. shown in miles
Distances in Canada shown in kilometers

0   20   40   60 Miles
0 20 40 60 80 100 Kilometers

© 1996 GeoSystems Global Corp.

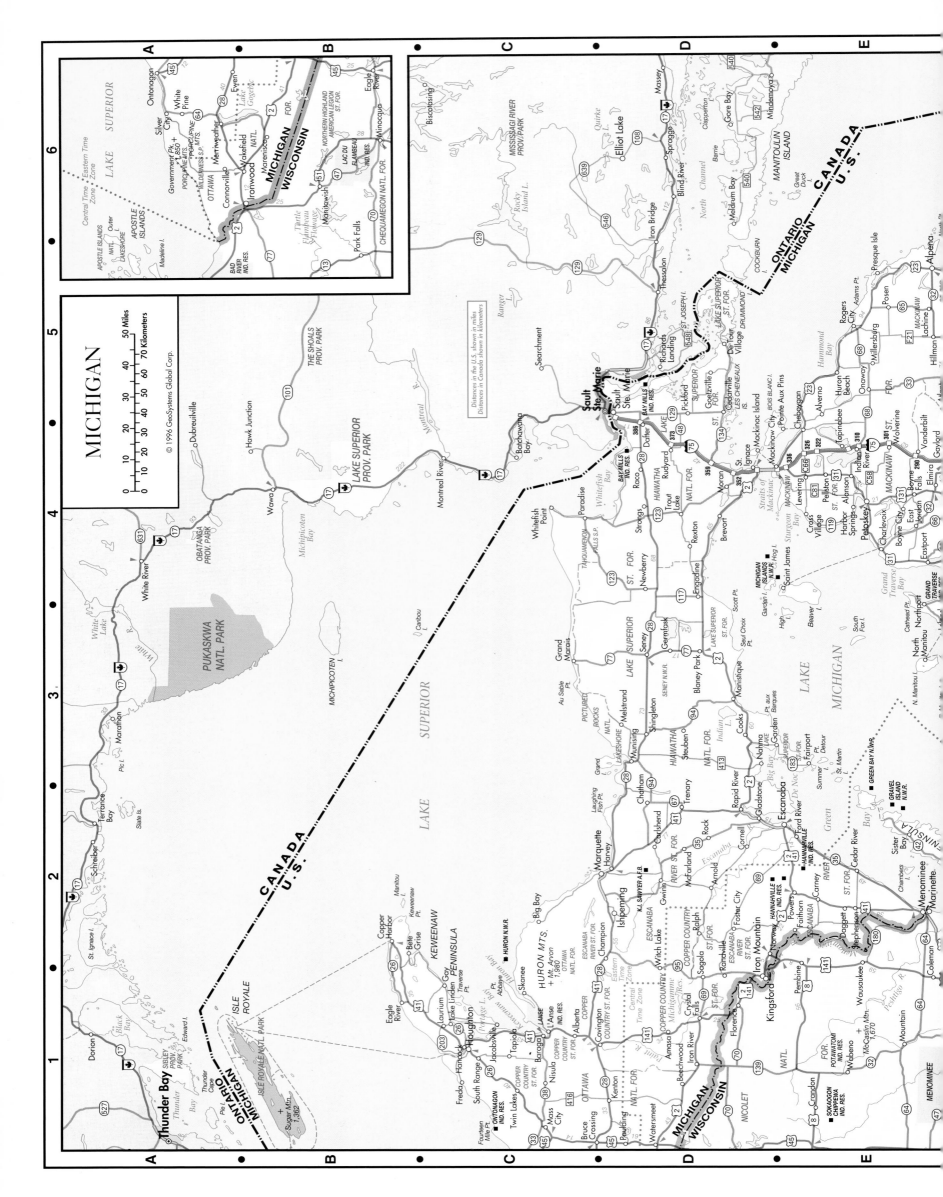

MICHIGAN

© 1996 GeoSystems Global Corp.

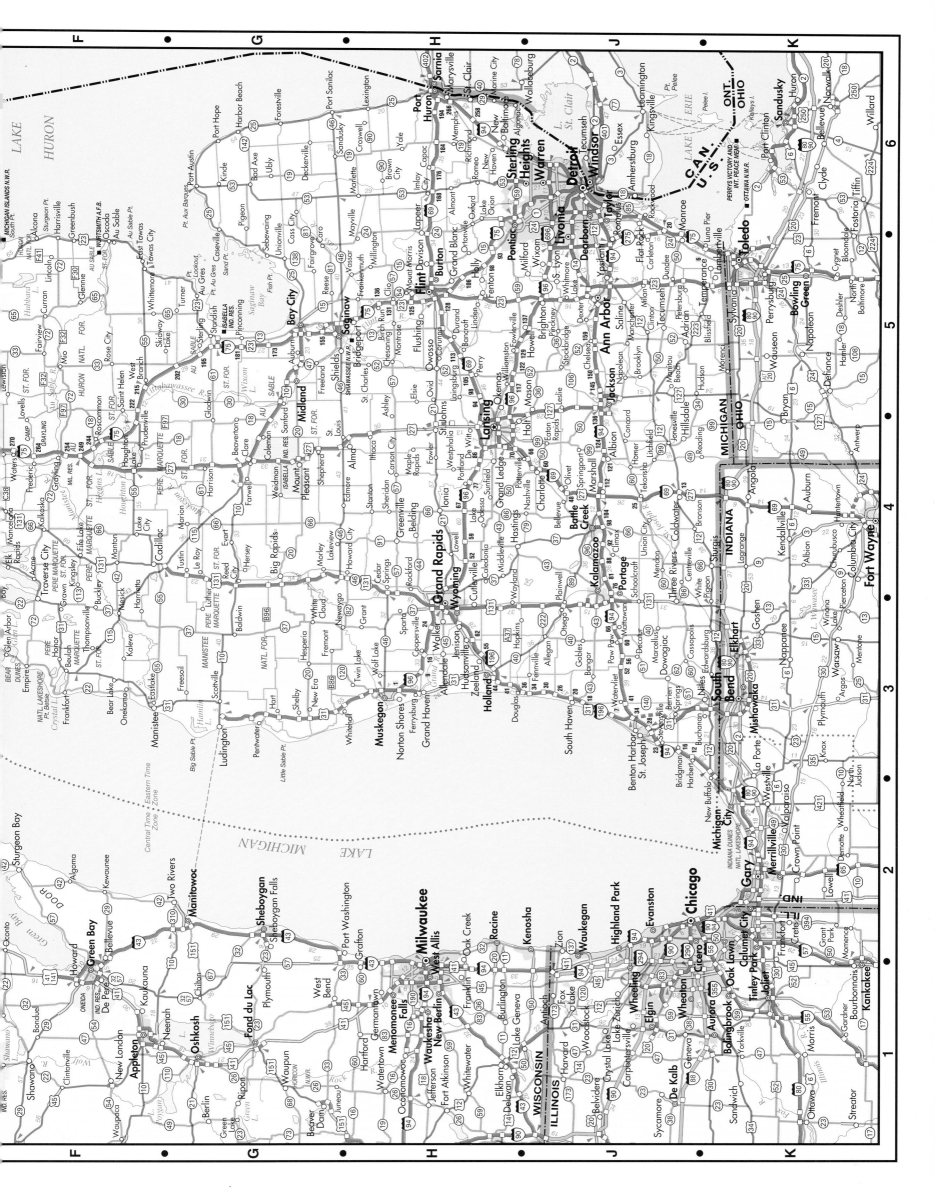

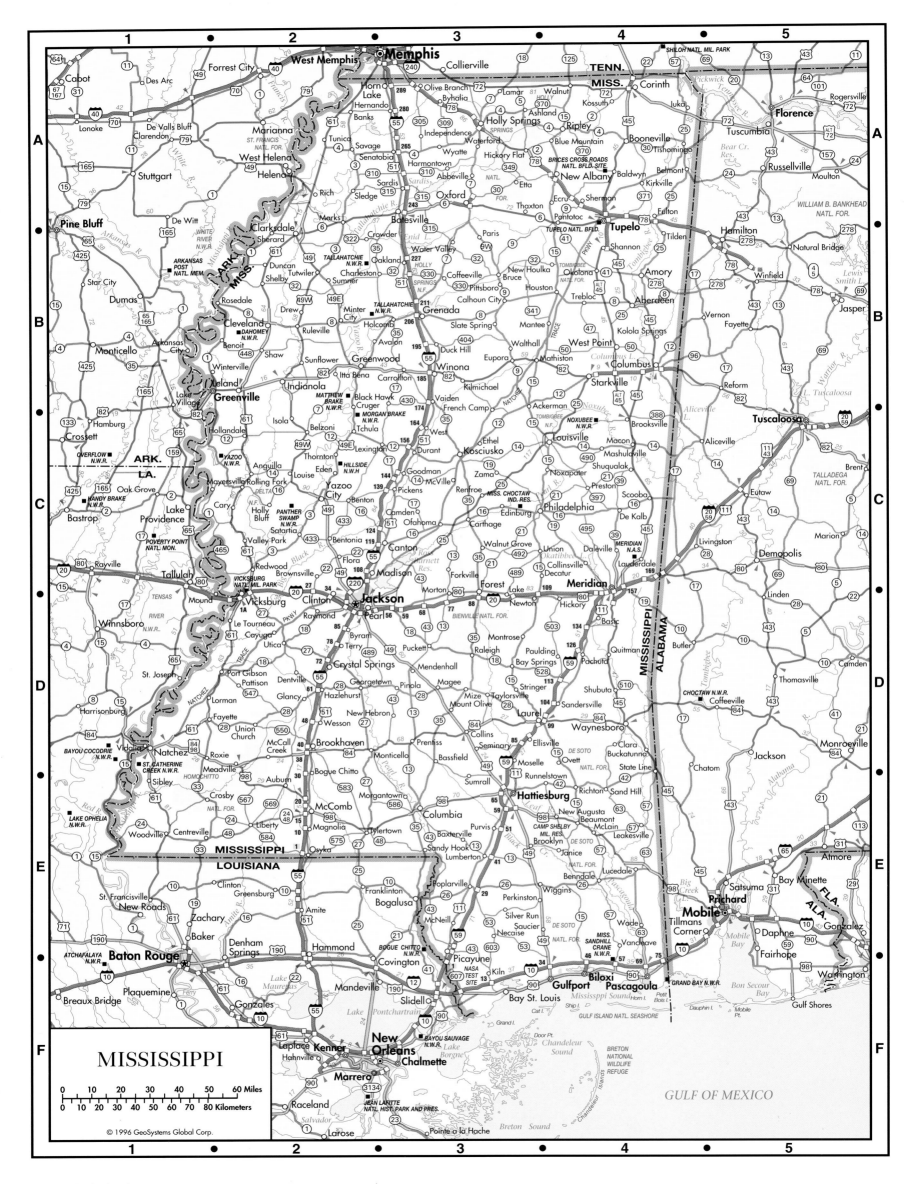

MISSISSIPPI

0 10 20 30 40 50 60 Miles
0 10 20 30 40 50 60 70 80 Kilometers

© 1996 GeoSystems Global Corp.

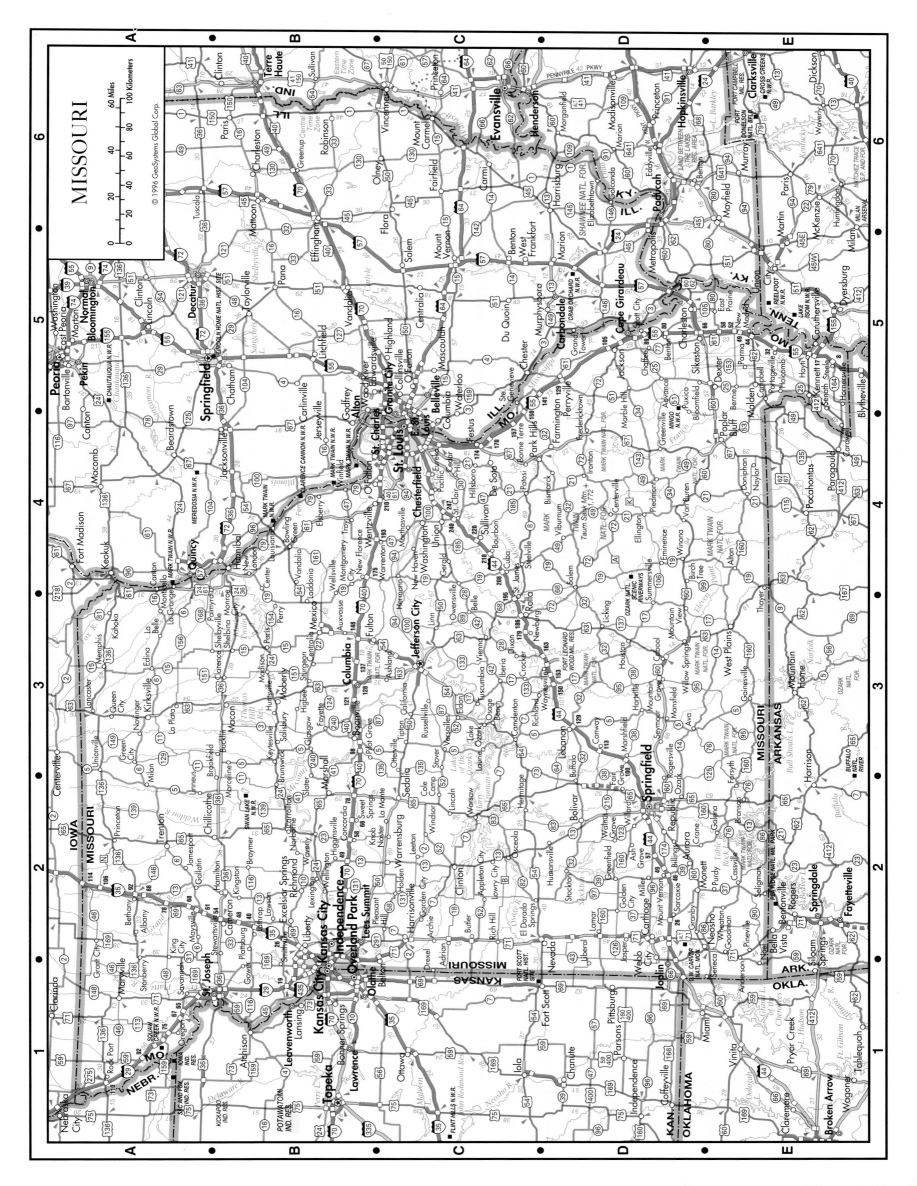

MISSOURI

© 1996 GeoSystems Global Corp.

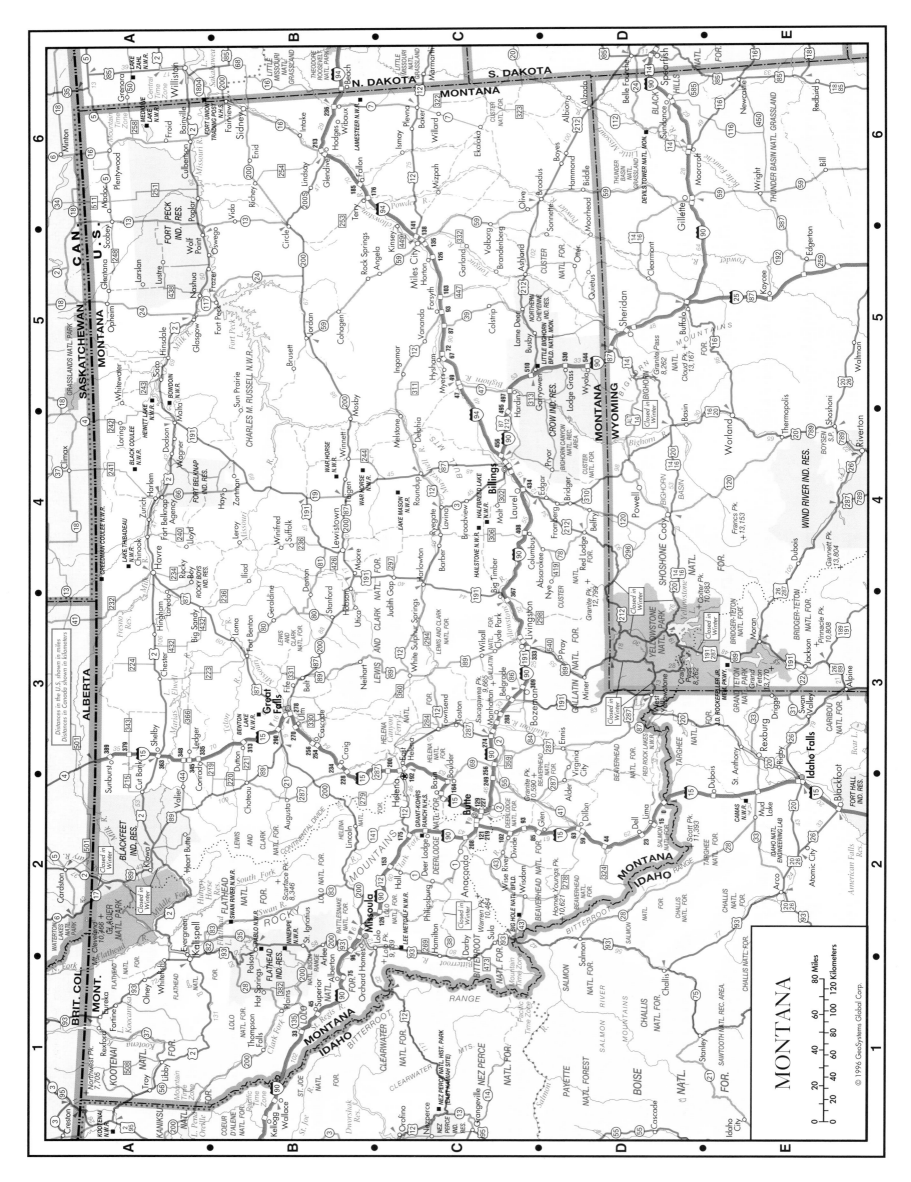

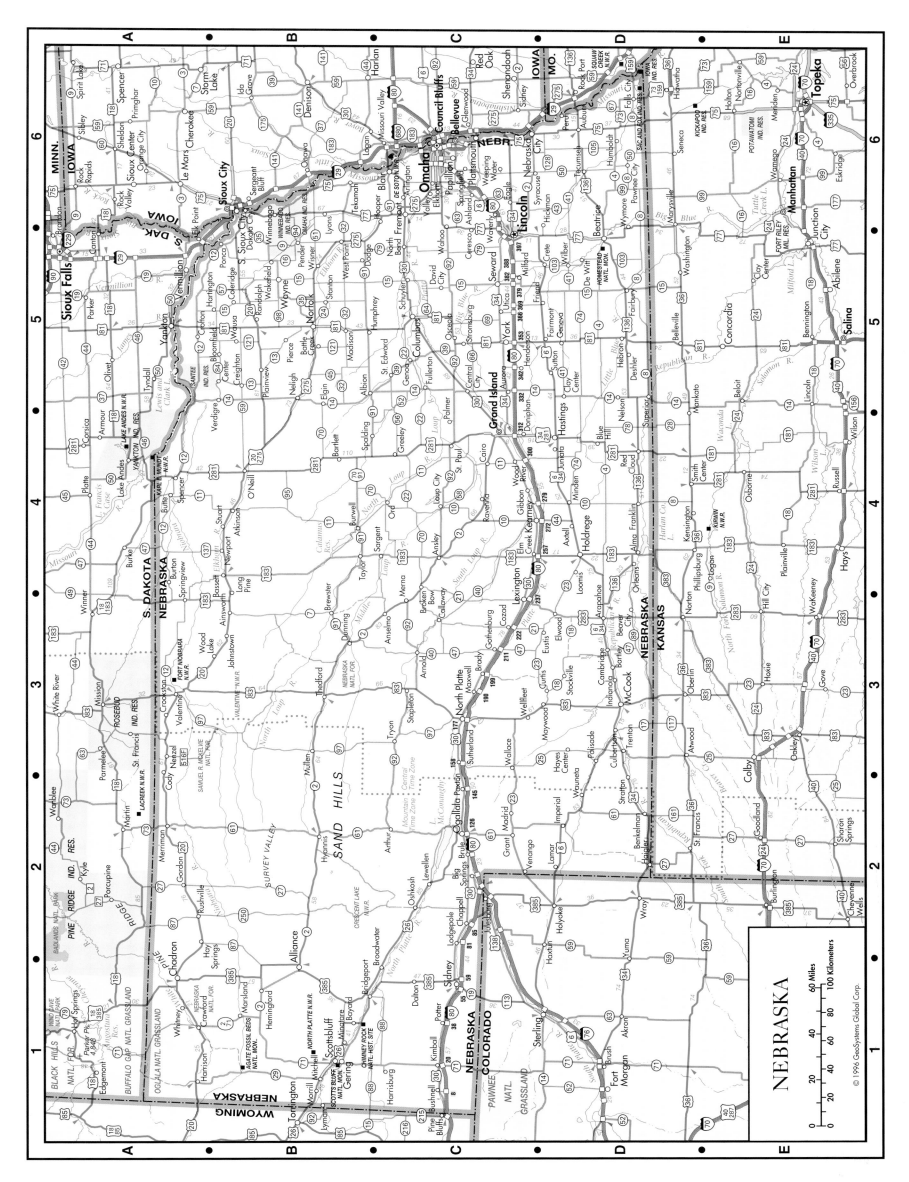

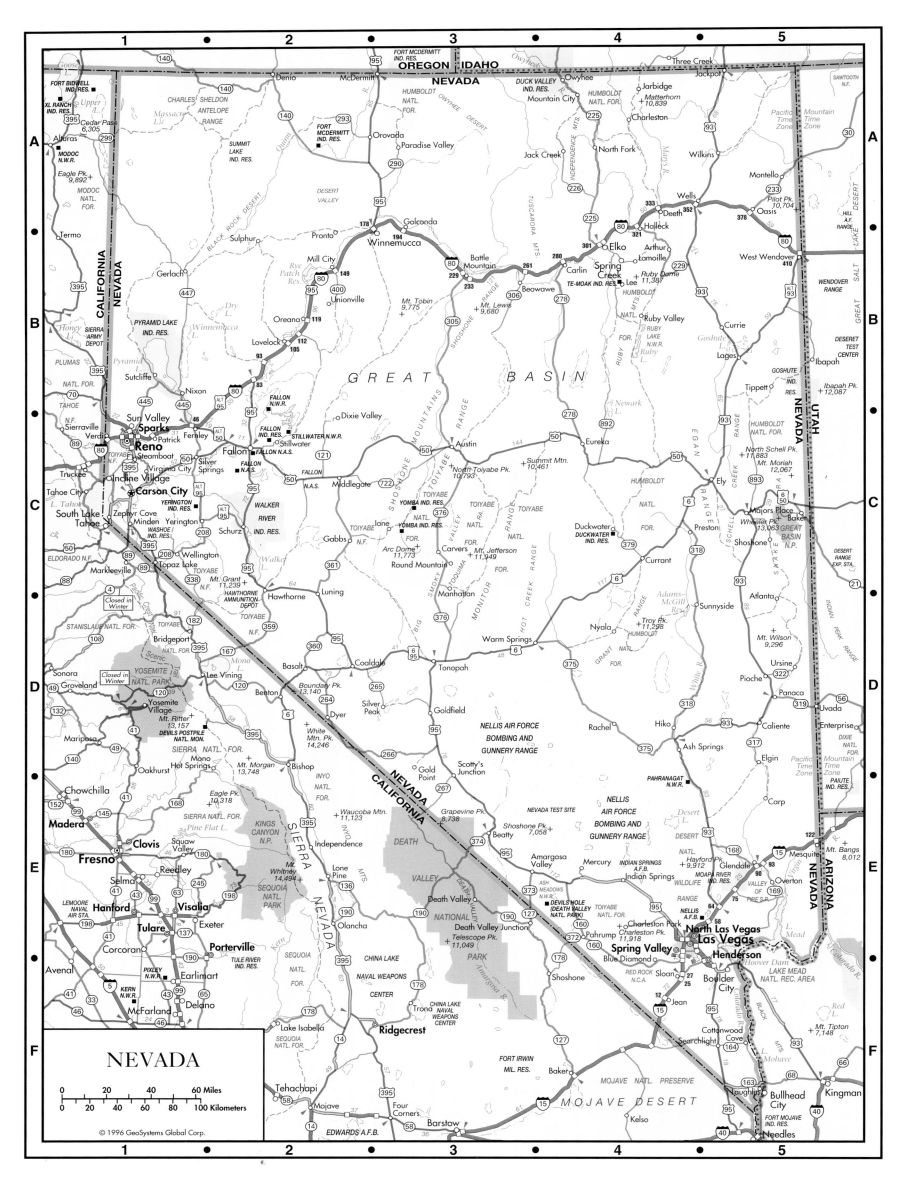

NEVADA

0   20   40   60 Miles
0   20   40   60   80   100 Kilometers

© 1996 GeoSystems Global Corp.

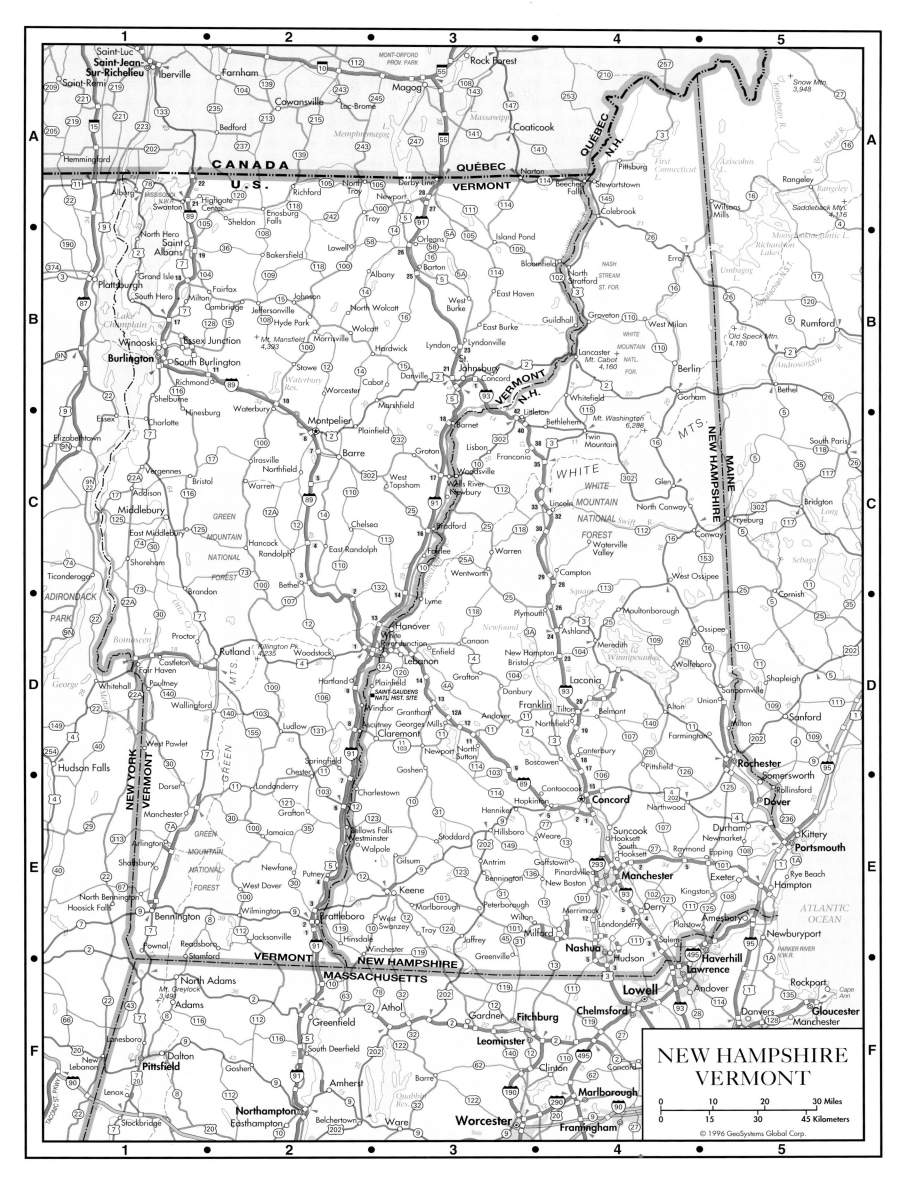

NEW HAMPSHIRE
VERMONT

0   10   20   30 Miles
0 15   30   45 Kilometers

© 1996 GeoSystems Global Corp.

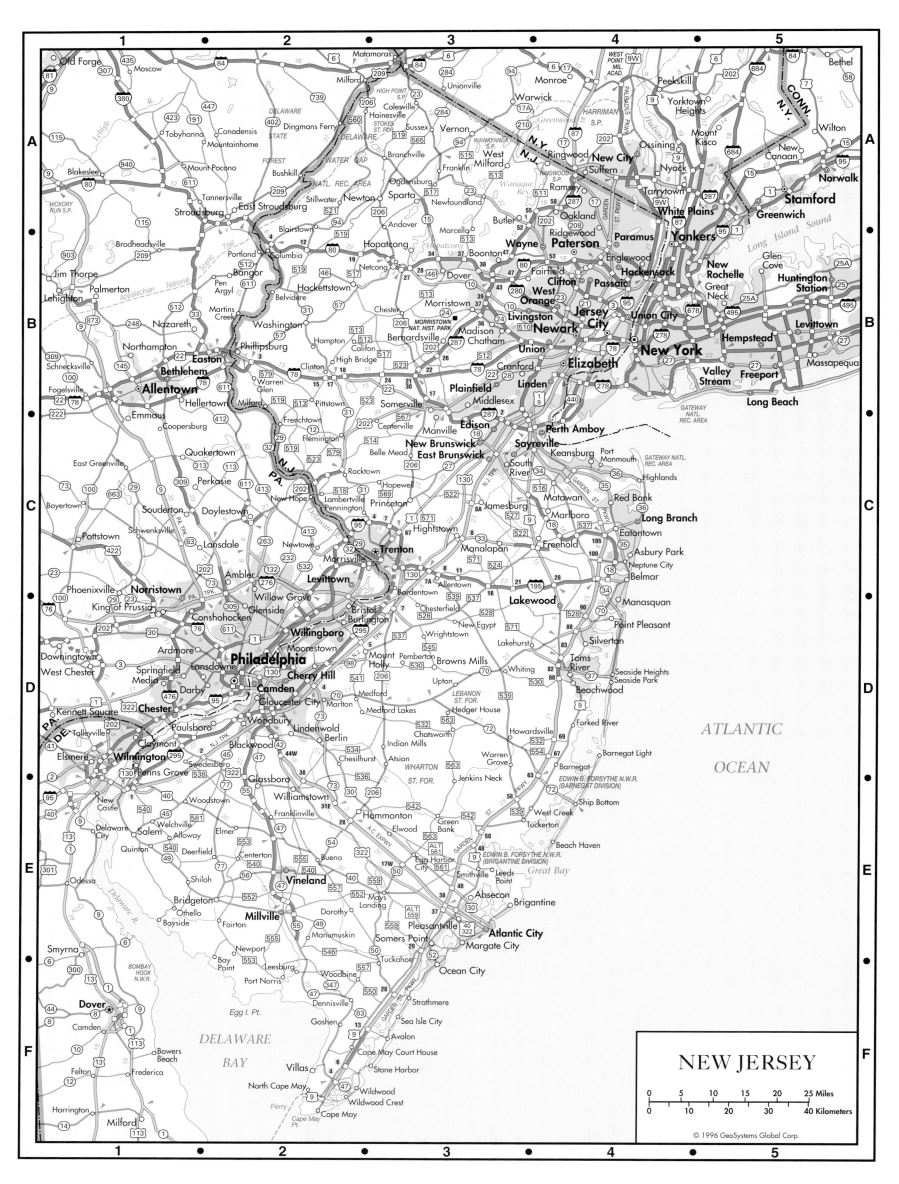

NEW JERSEY

| 0 | 5 | 10 | 15 | 20 | 25 Miles |

| 0 | 10 | 20 | 30 | 40 Kilometers |

© 1996 GeoSystems Global Corp.

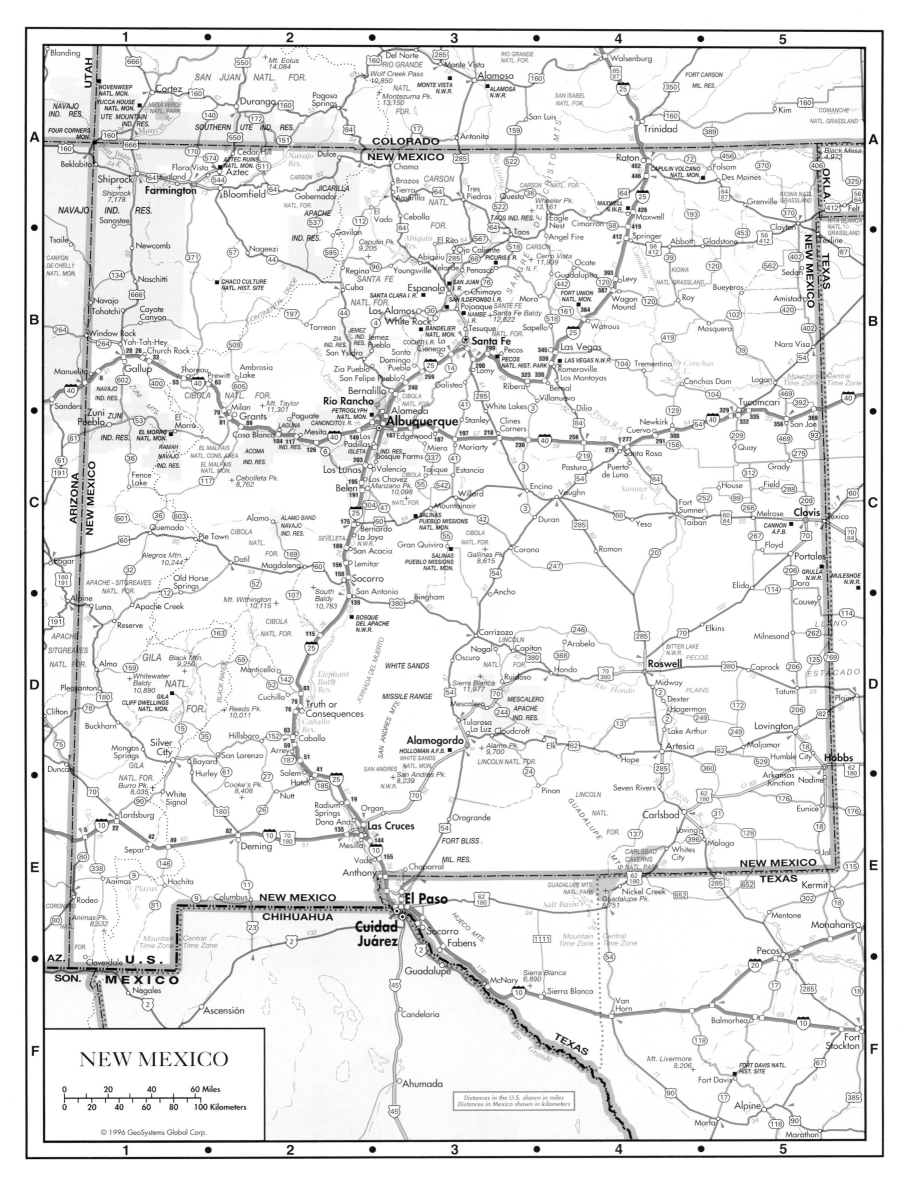

NEW MEXICO

0 20 40 60 Miles
0 20 40 60 80 100 Kilometers

© 1996 GeoSystems Global Corp.

Distances in the U.S. shown in miles
Distances in Mexico shown in kilometers

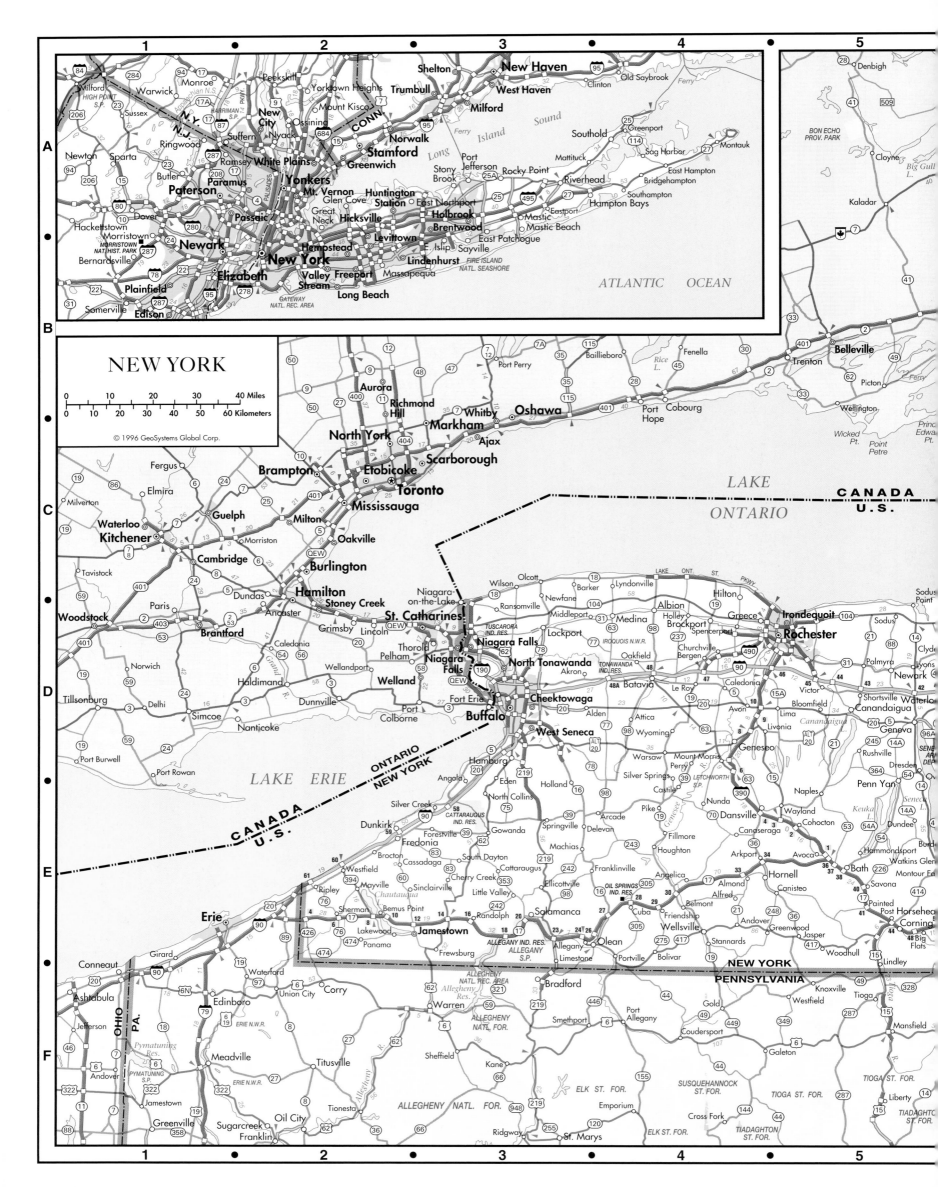

NEW YORK

© 1996 GeoSystems Global Corp.

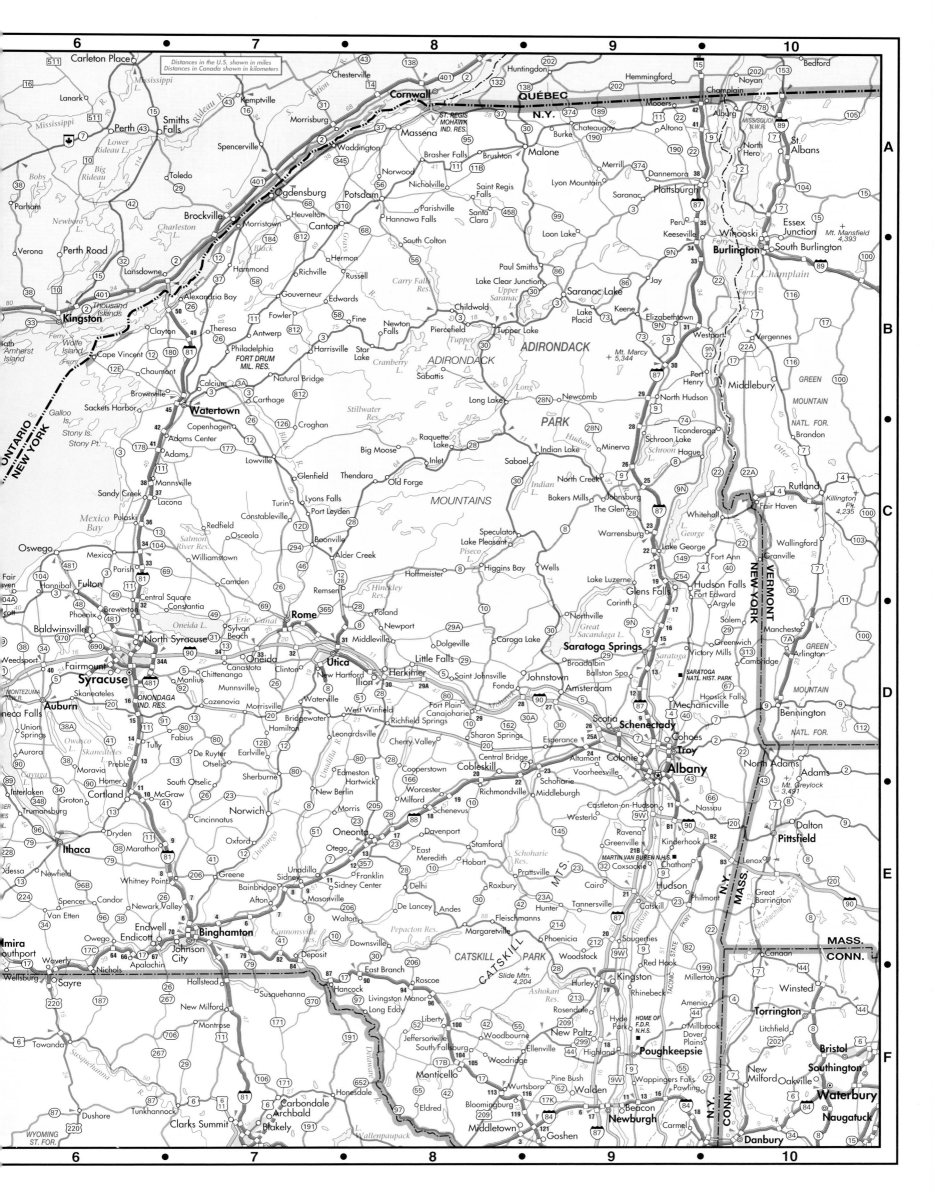

New York 135

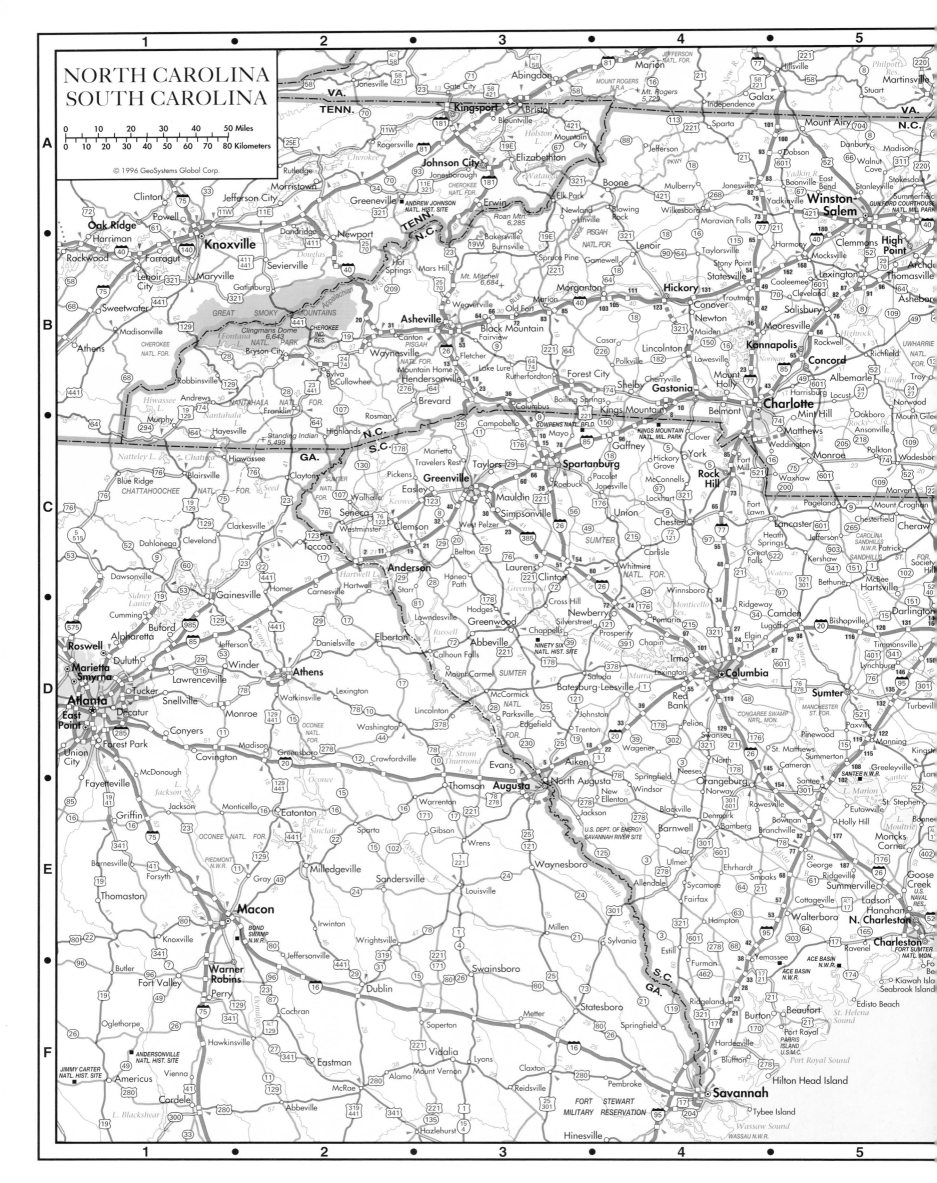

NORTH CAROLINA
SOUTH CAROLINA

© 1996 GeoSystems Global Corp.

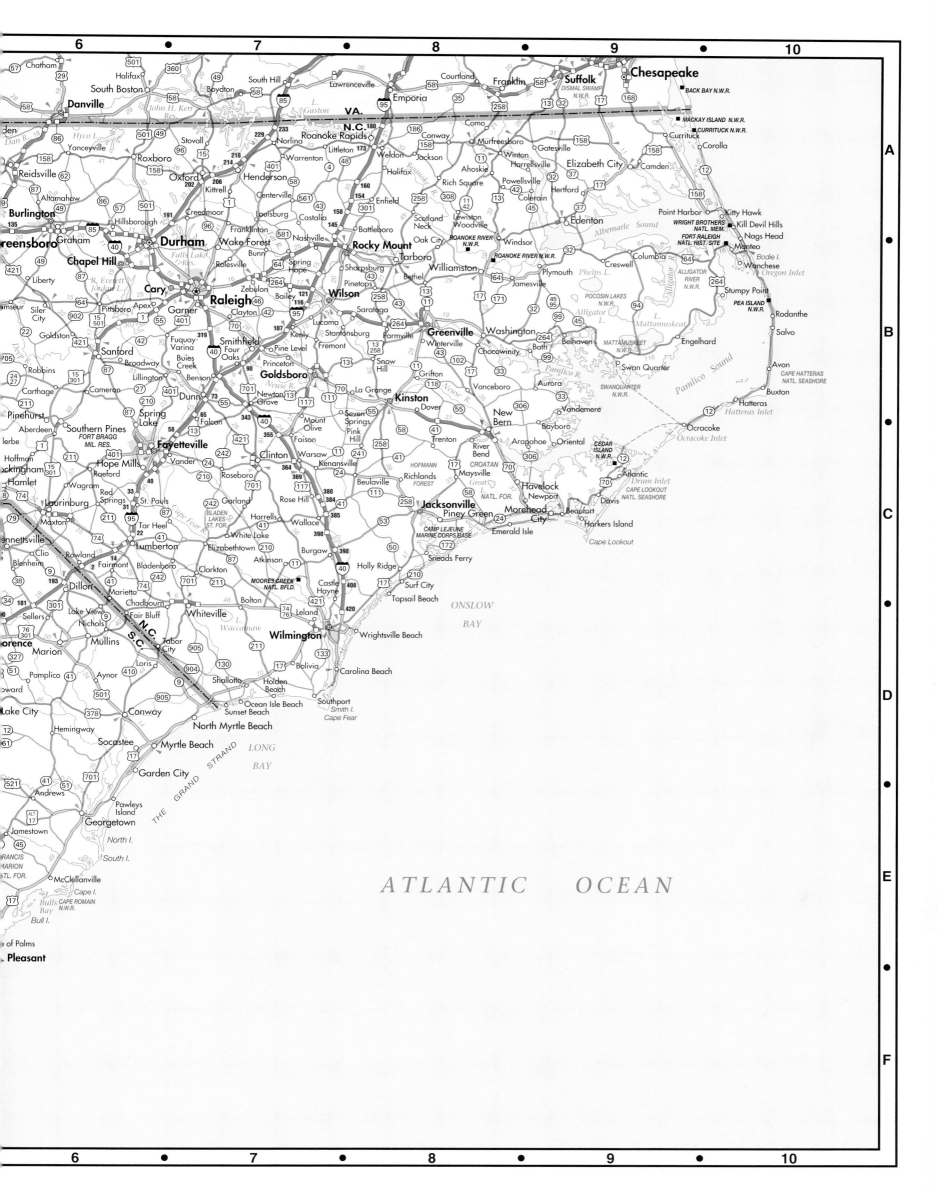

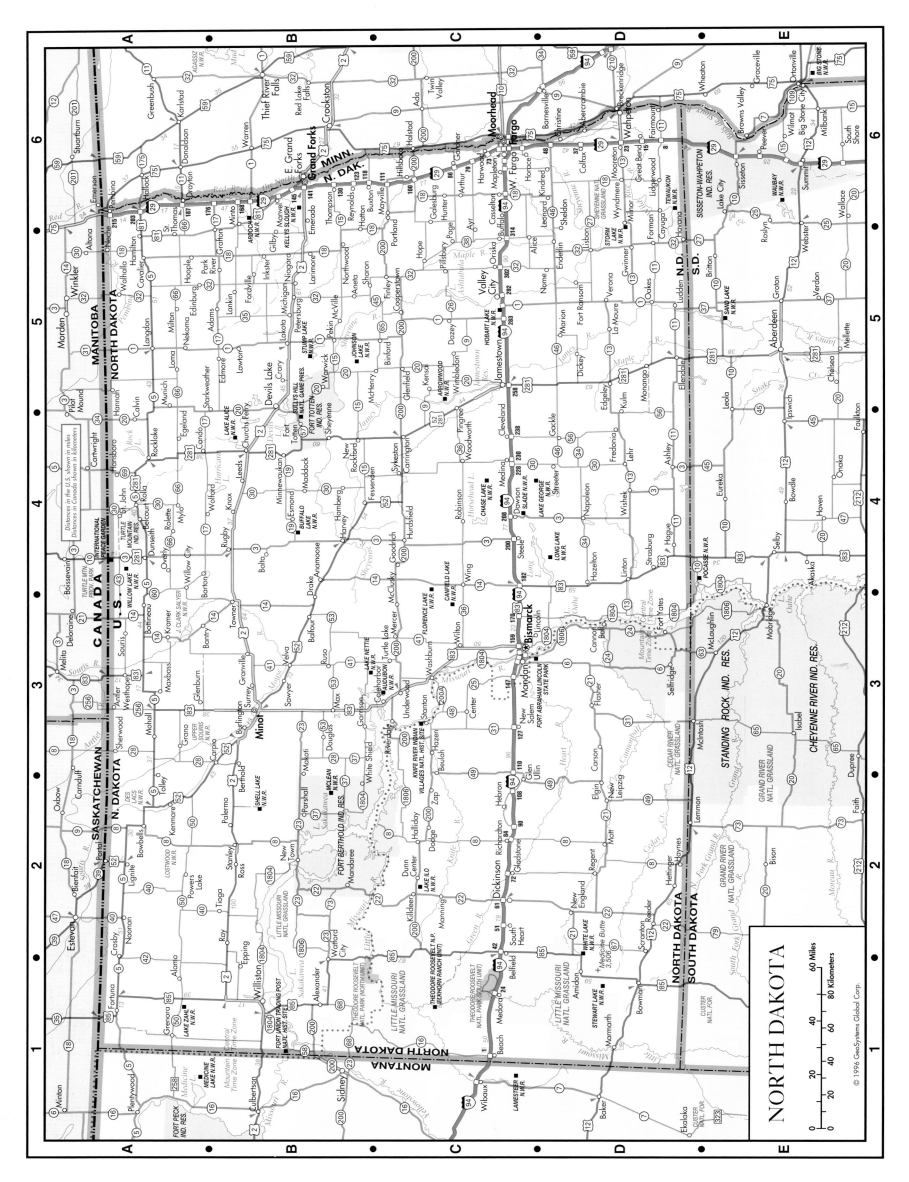

NORTH DAKOTA

© 1996 GeoSystems Global Corp.

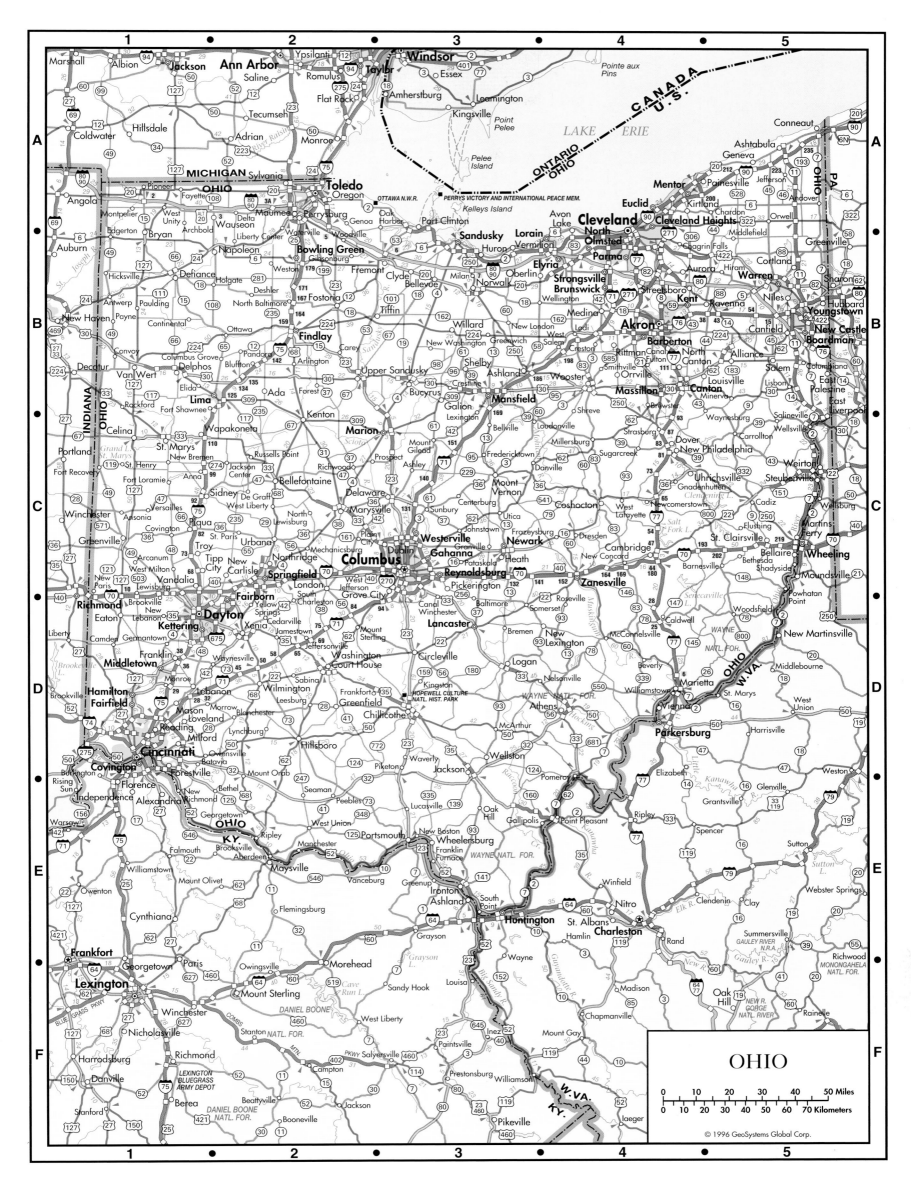

OHIO

© 1996 GeoSystems Global Corp.

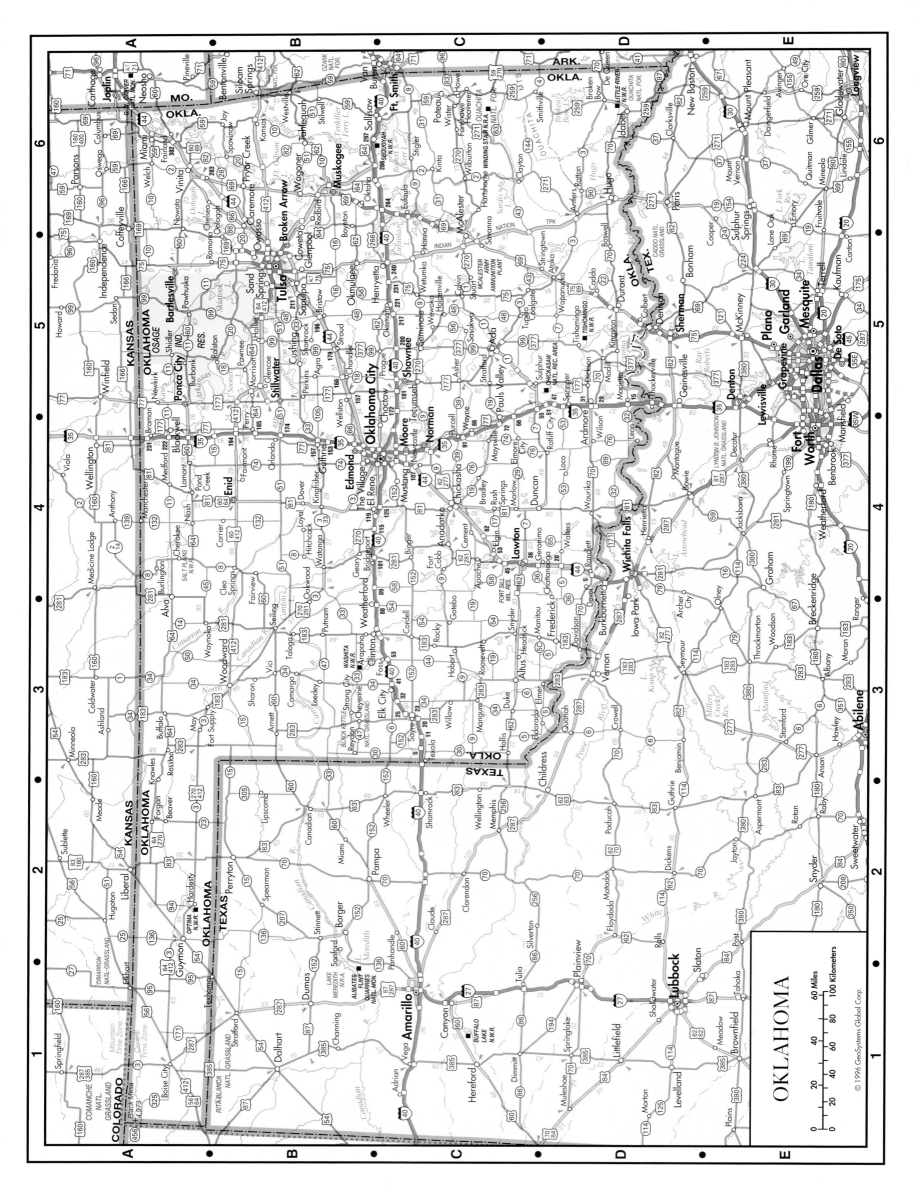

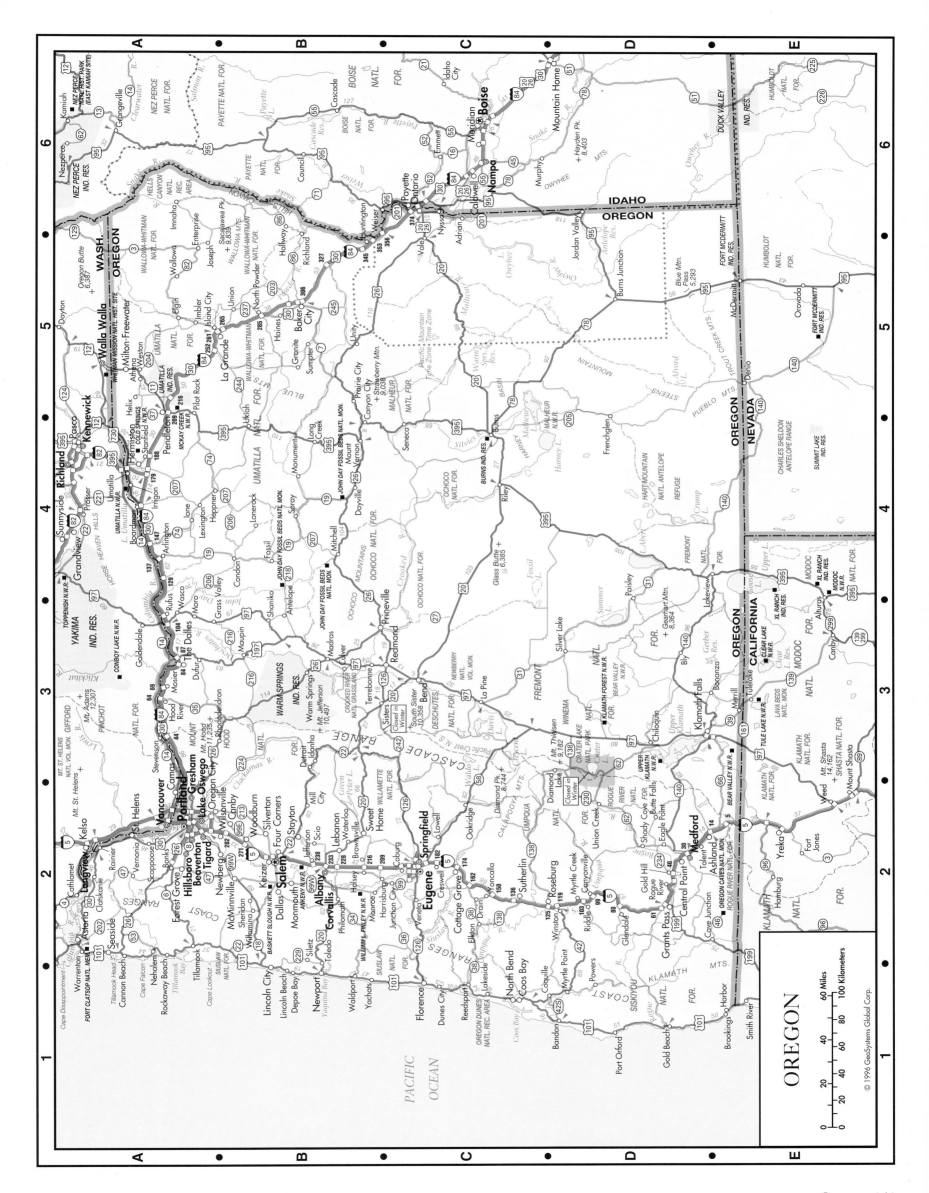

OREGON

© 1996 GeoSystems Global Corp.

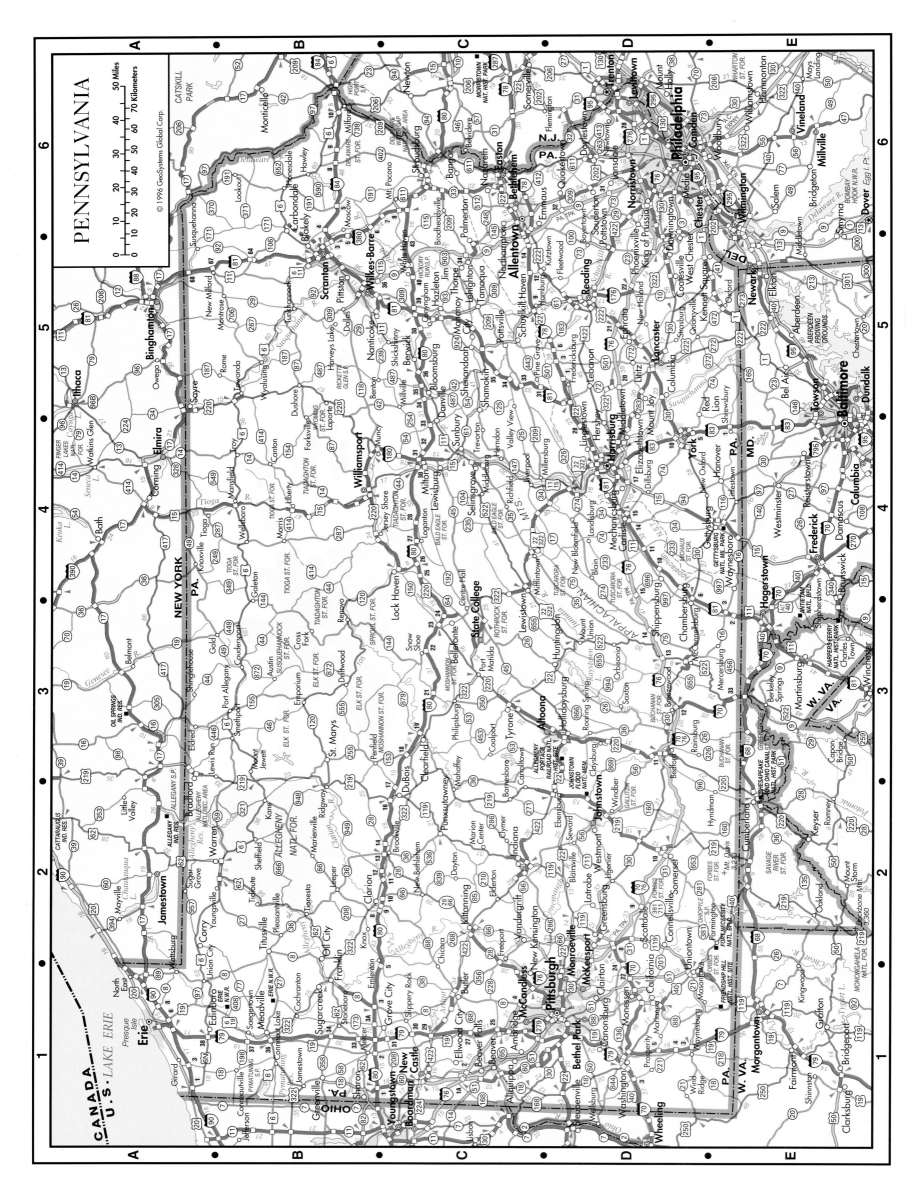

PENNSYLVANIA

© 1996 GeoSystems Global Corp.

50 Miles
70 Kilometers

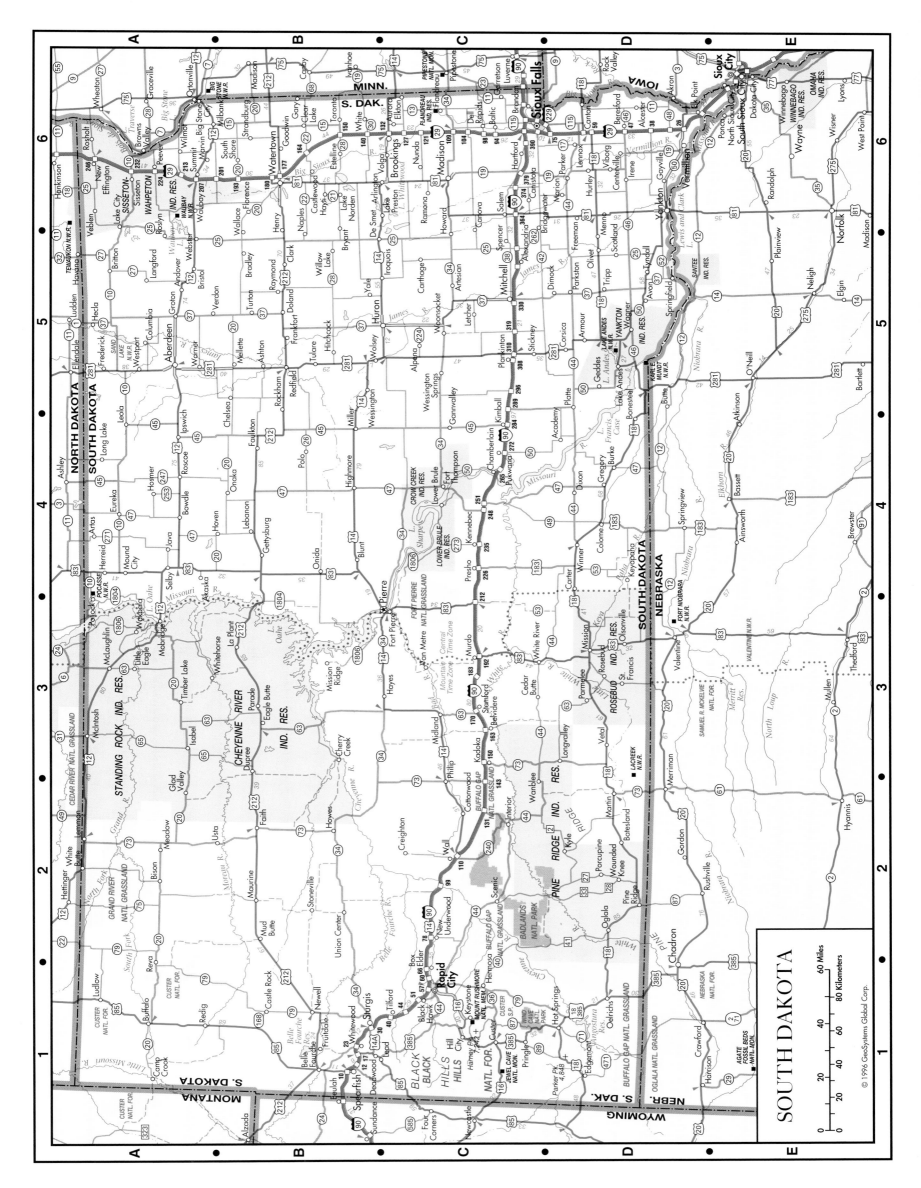

South Dakota 143

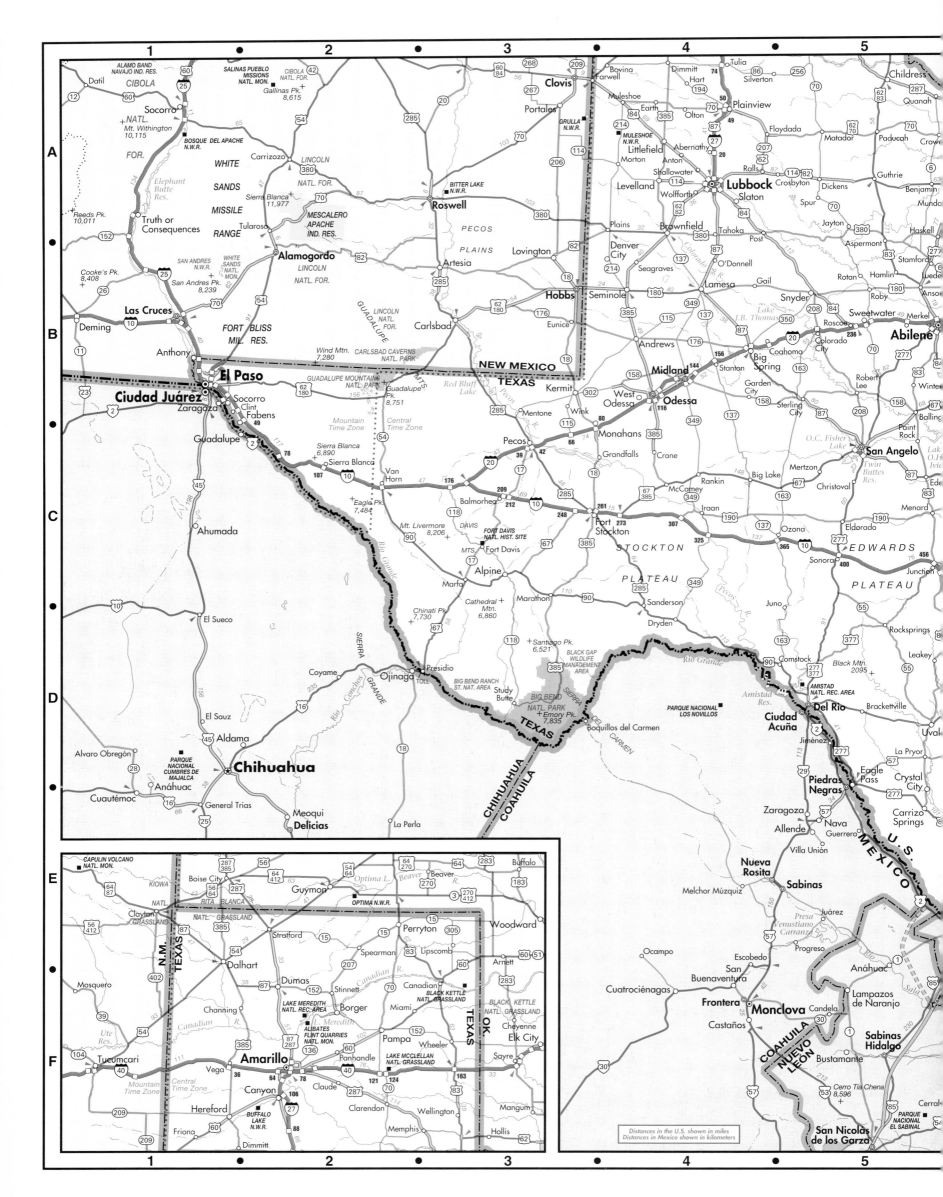

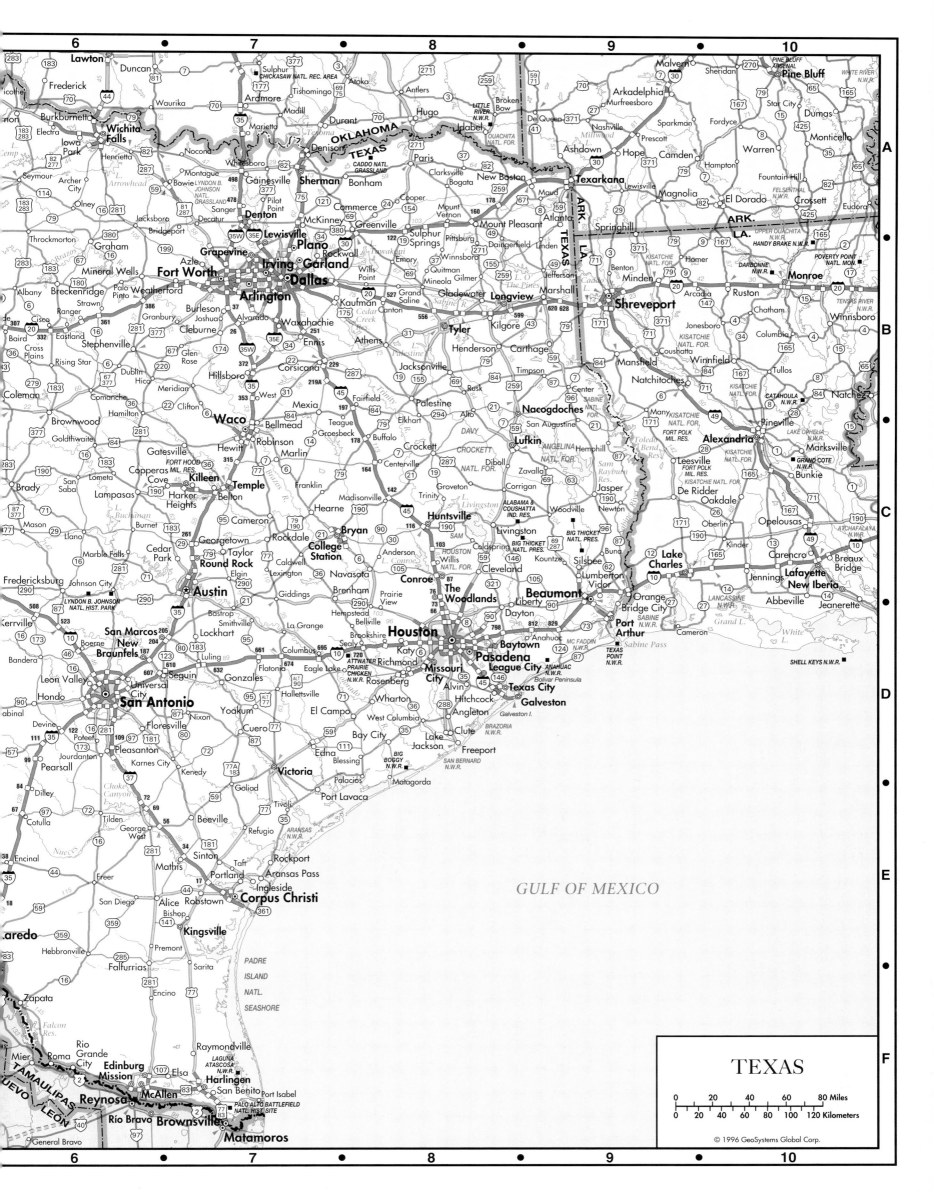

GULF OF MEXICO

TEXAS

0   20   40   60   80 Miles
0 20 40 60 80 100 120 Kilometers

© 1996 GeoSystems Global Corp.

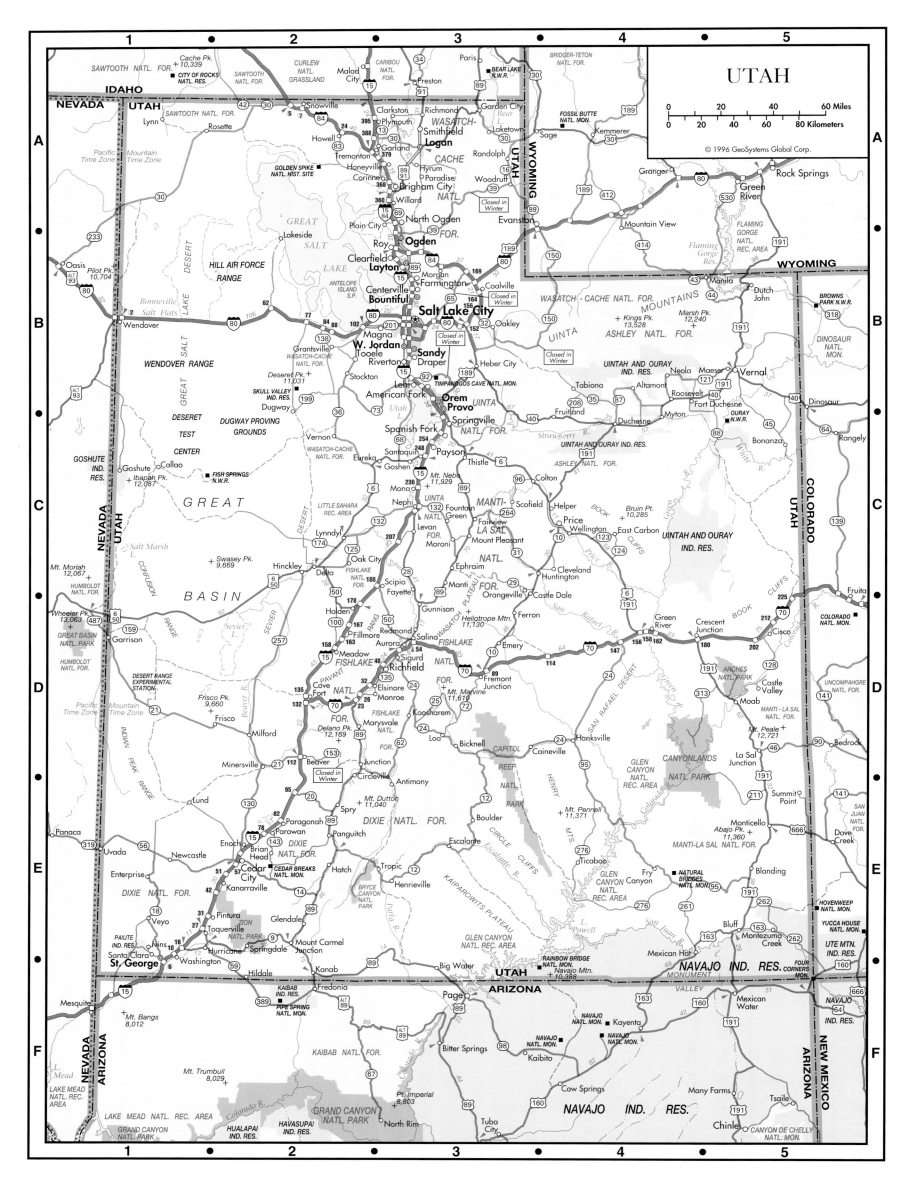

UTAH

© 1996 GeoSystems Global Corp.

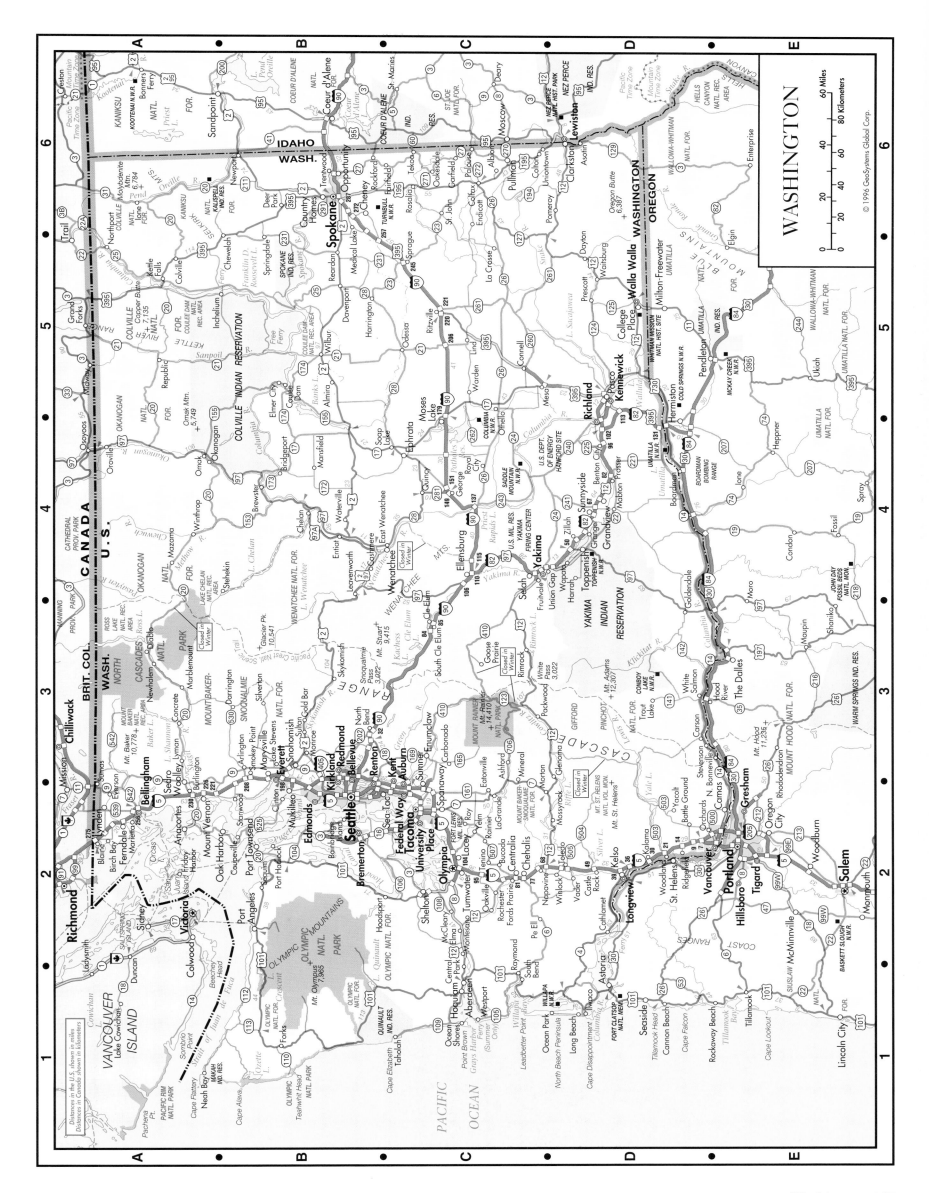

WASHINGTON

© 1996 GeoSystems Global Corp.

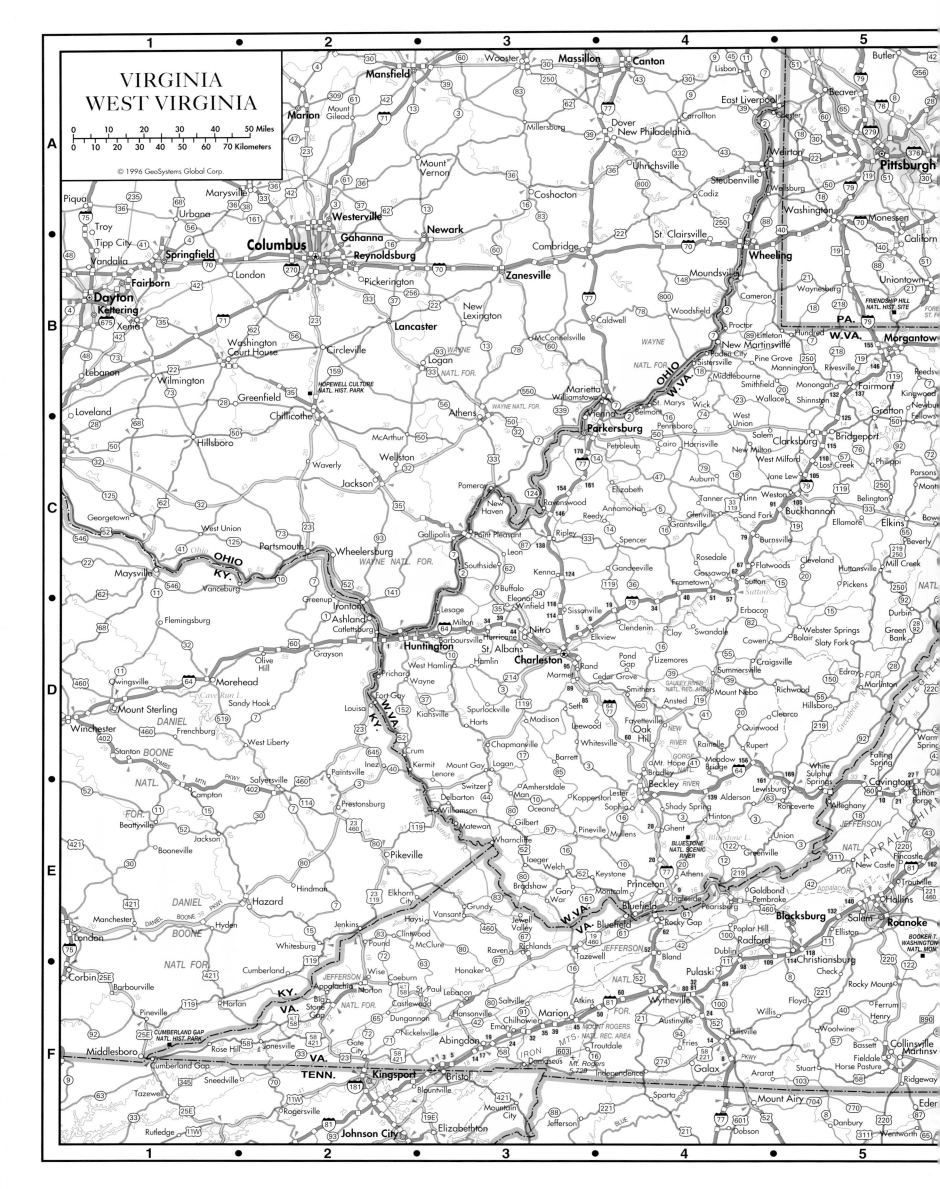

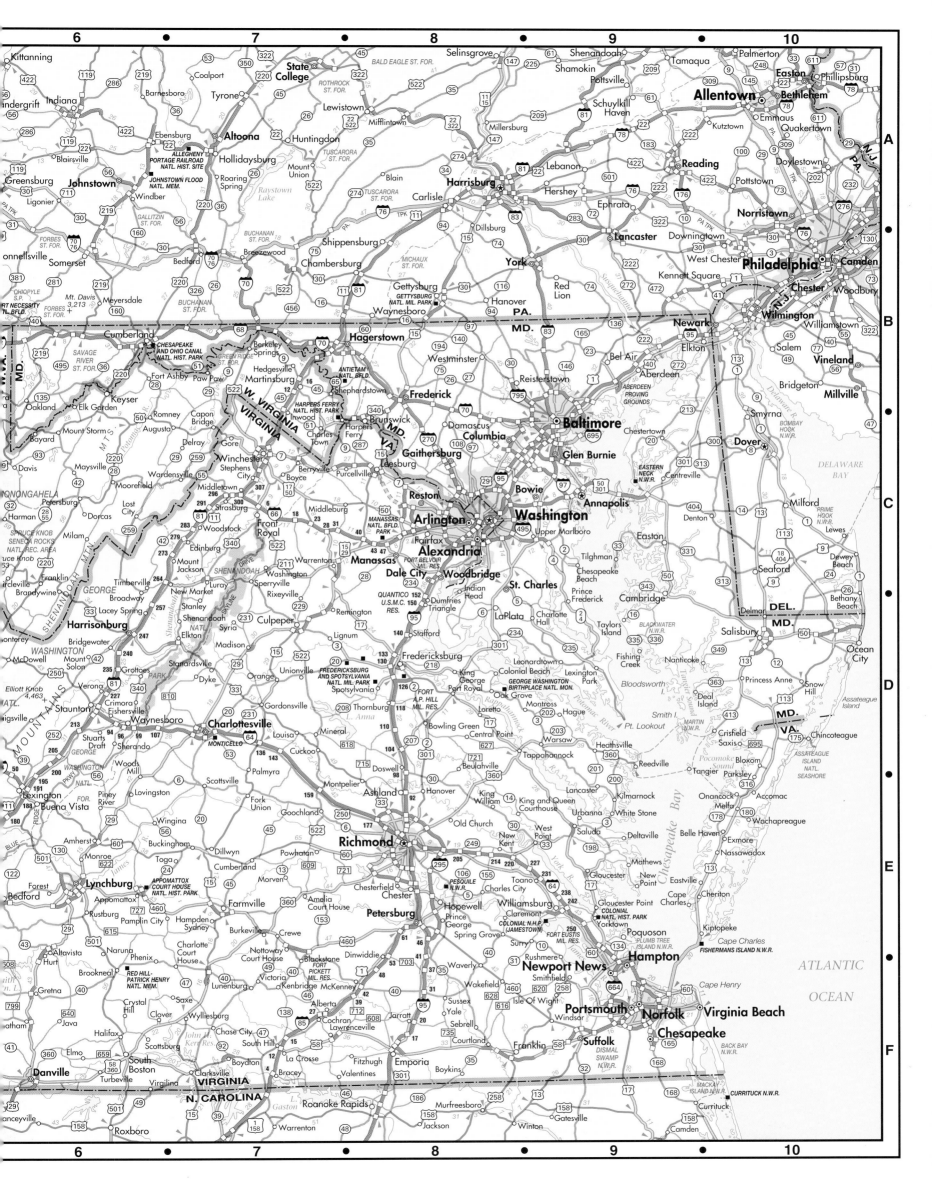

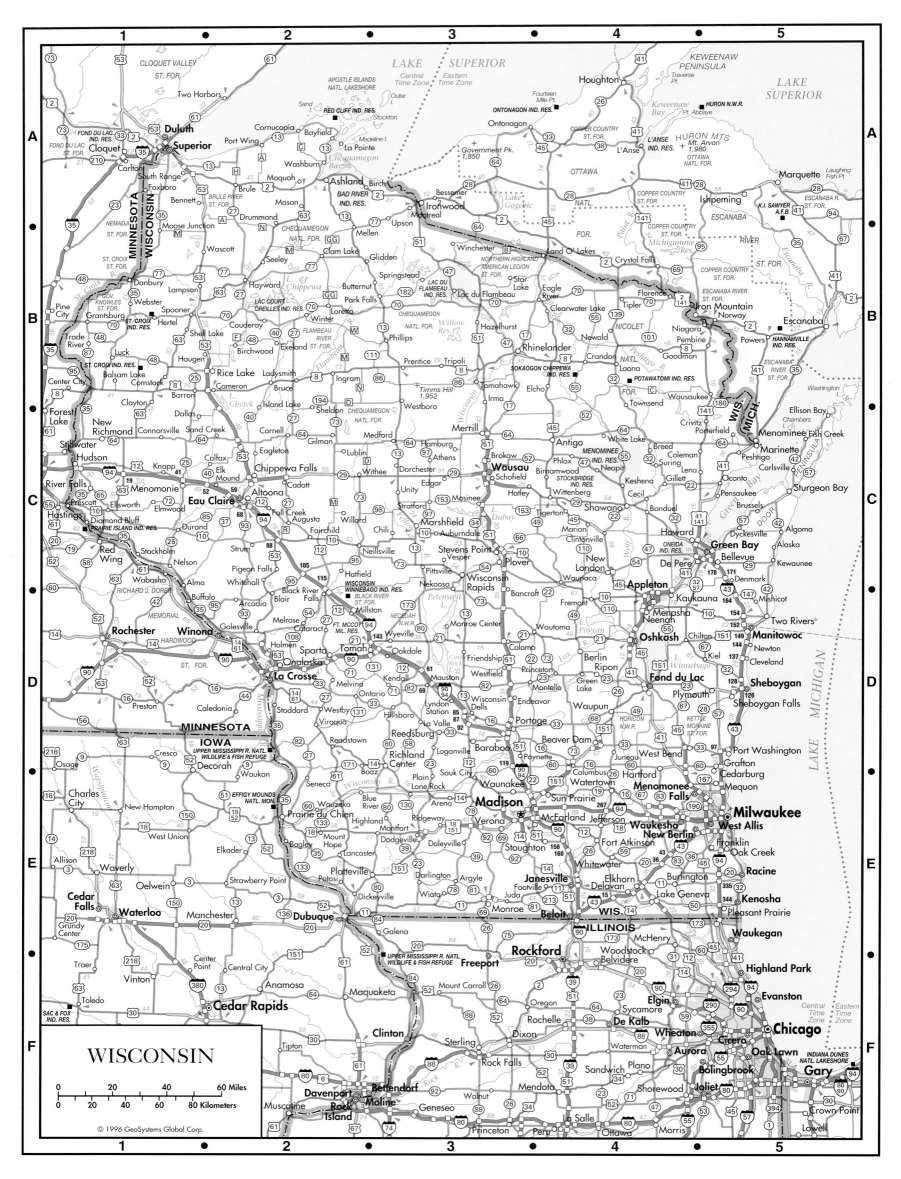

WISCONSIN

0   20   40   60 Miles
0 20 40 60 80 Kilometers

© 1996 GeoSystems Global Corp.

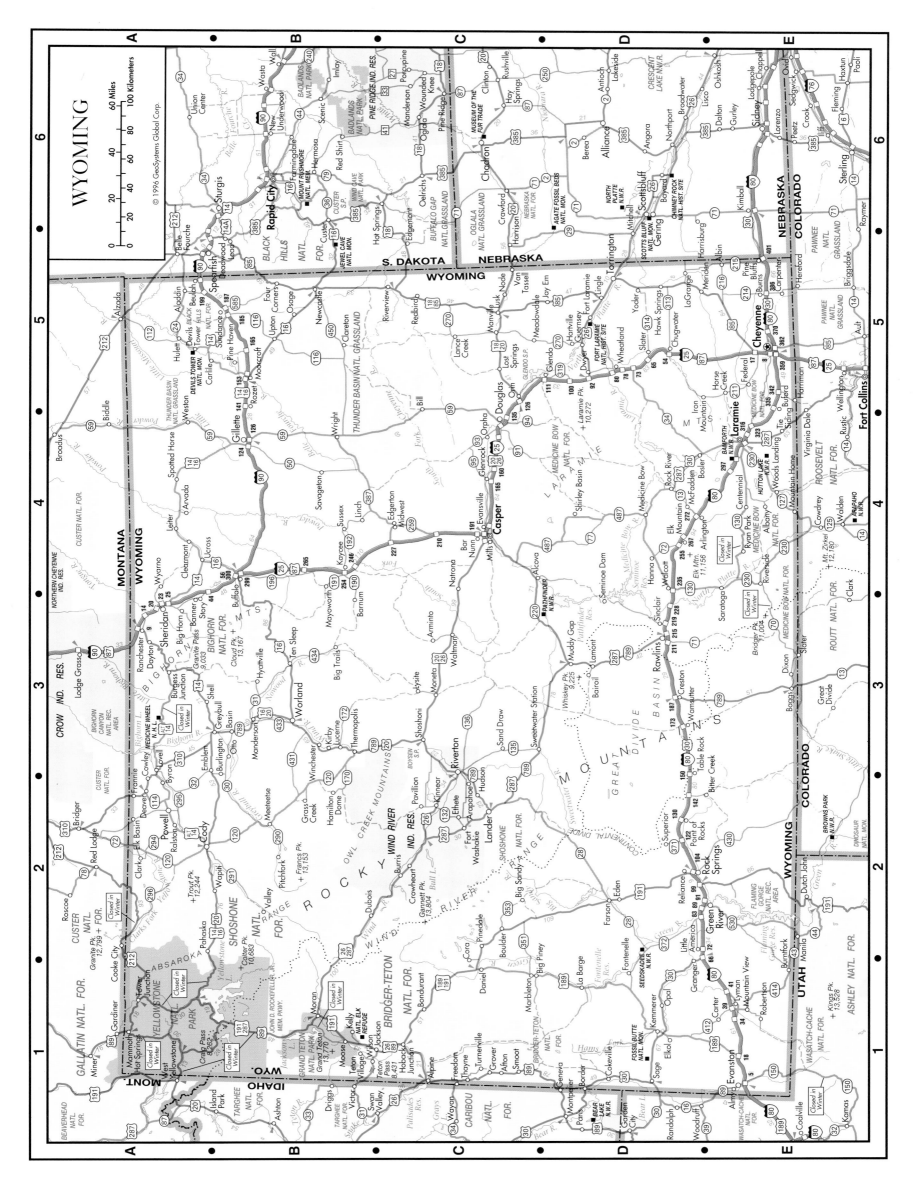

WYOMING

© 1996 GeoSystems Global Corp.

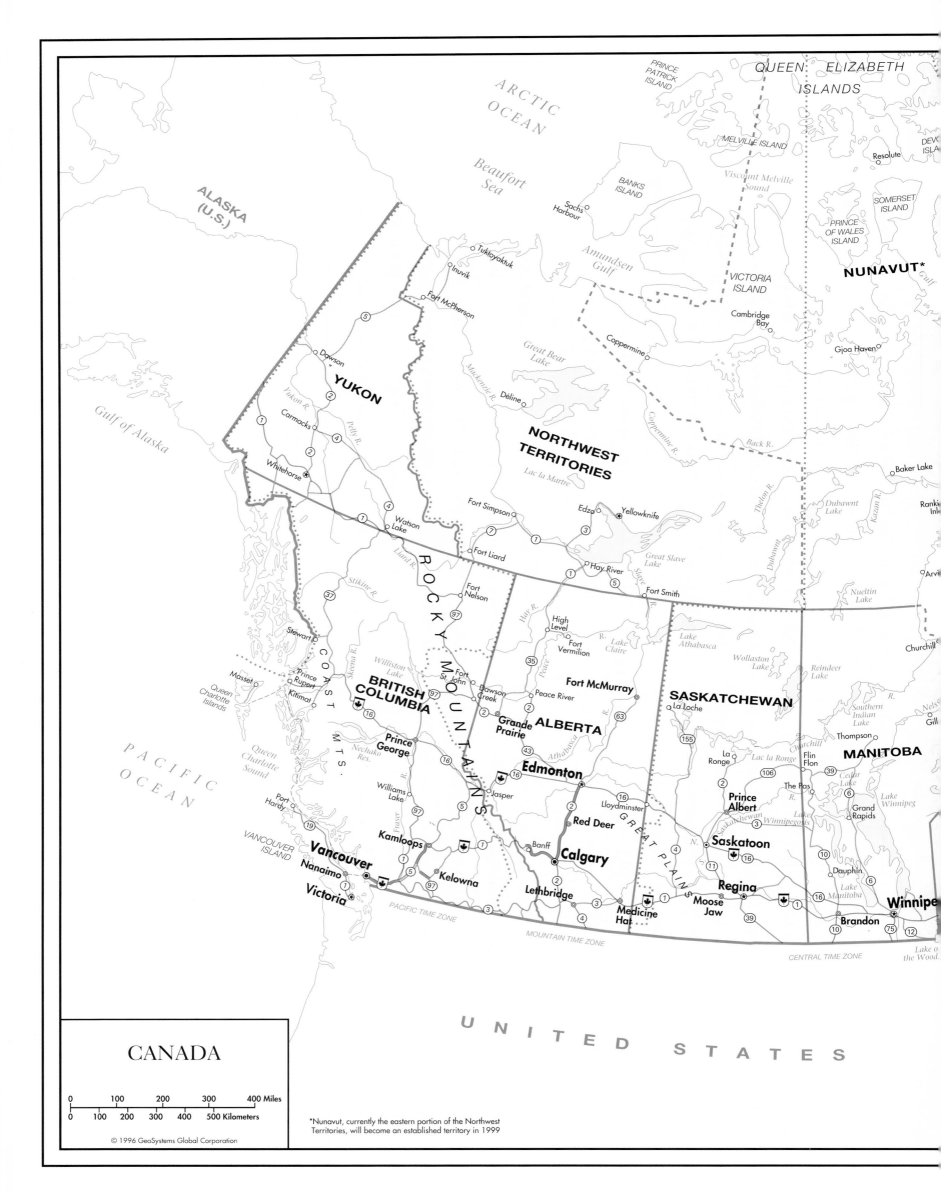

CANADA

0	100	200	300	400 Miles	
0	100	200	300	400	500 Kilometers

© 1996 GeoSystems Global Corporation

*Nunavut, currently the eastern portion of the Northwest
Territories, will become an established territory in 1999

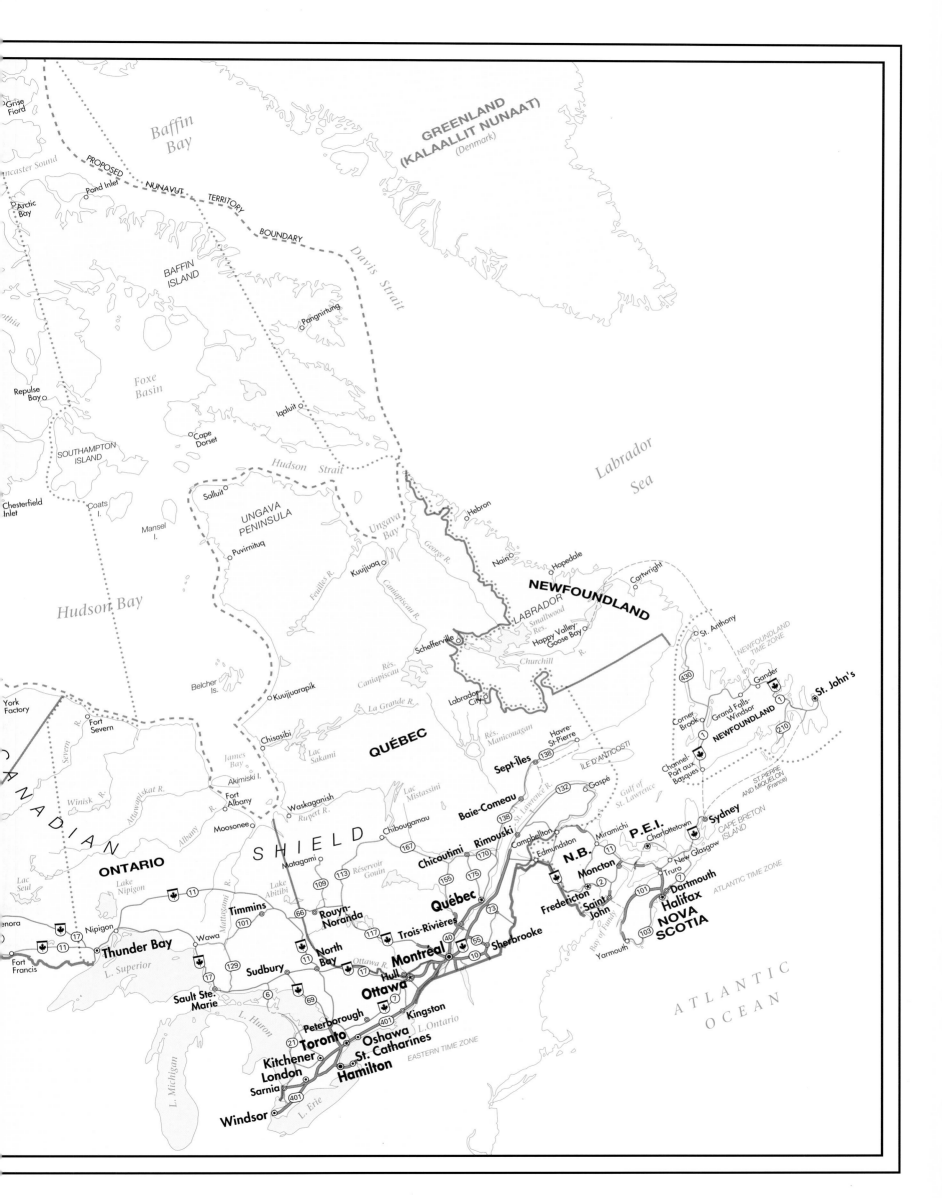

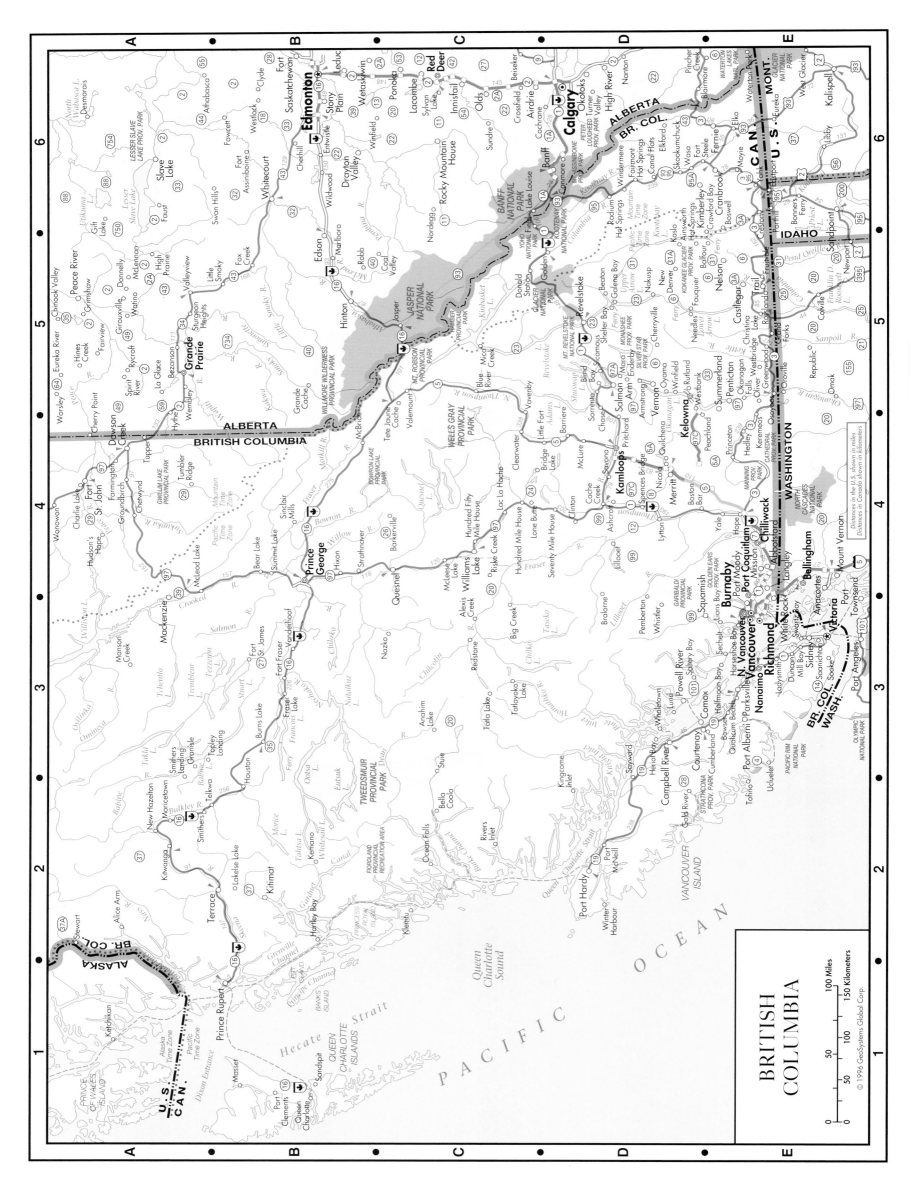

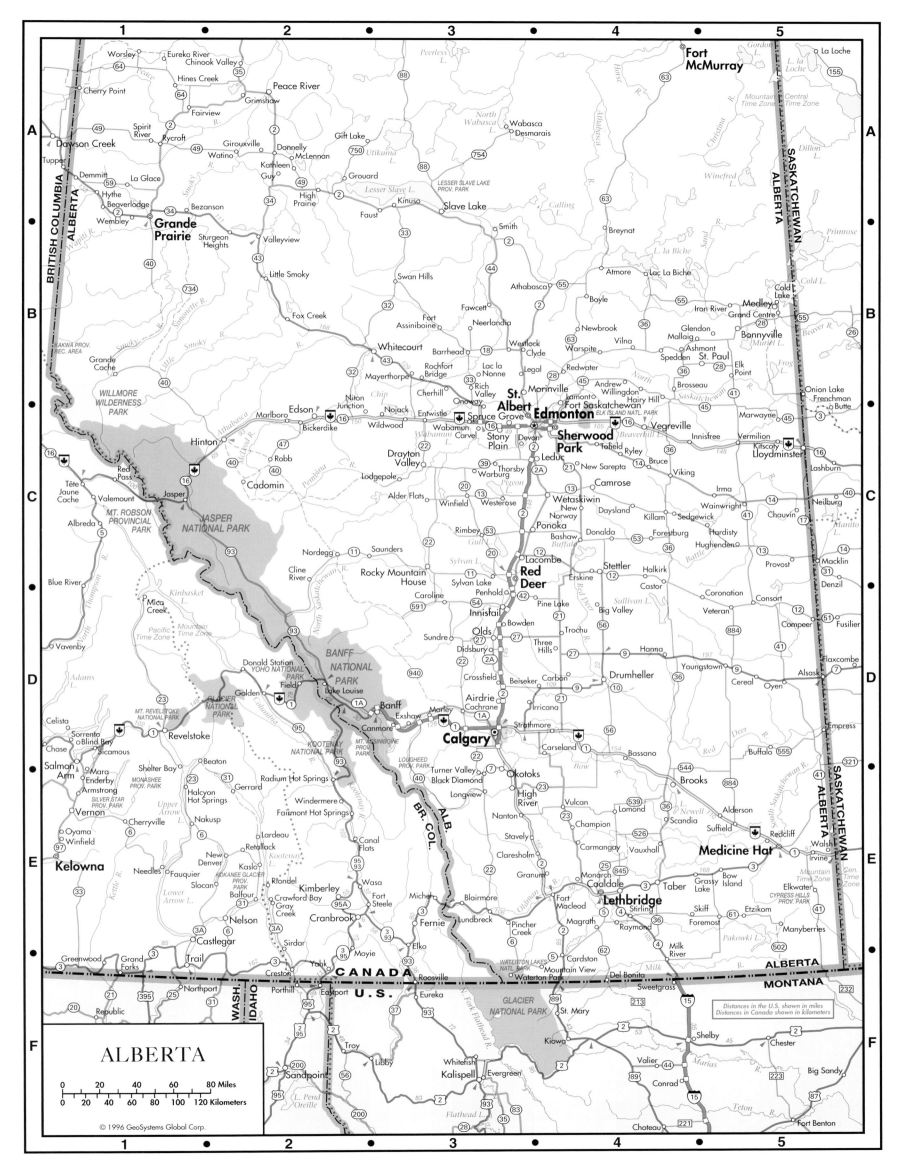

ALBERTA

0   20   40   60   80 Miles
0 20 40 60 80 100 120 Kilometers

© 1996 GeoSystems Global Corp.

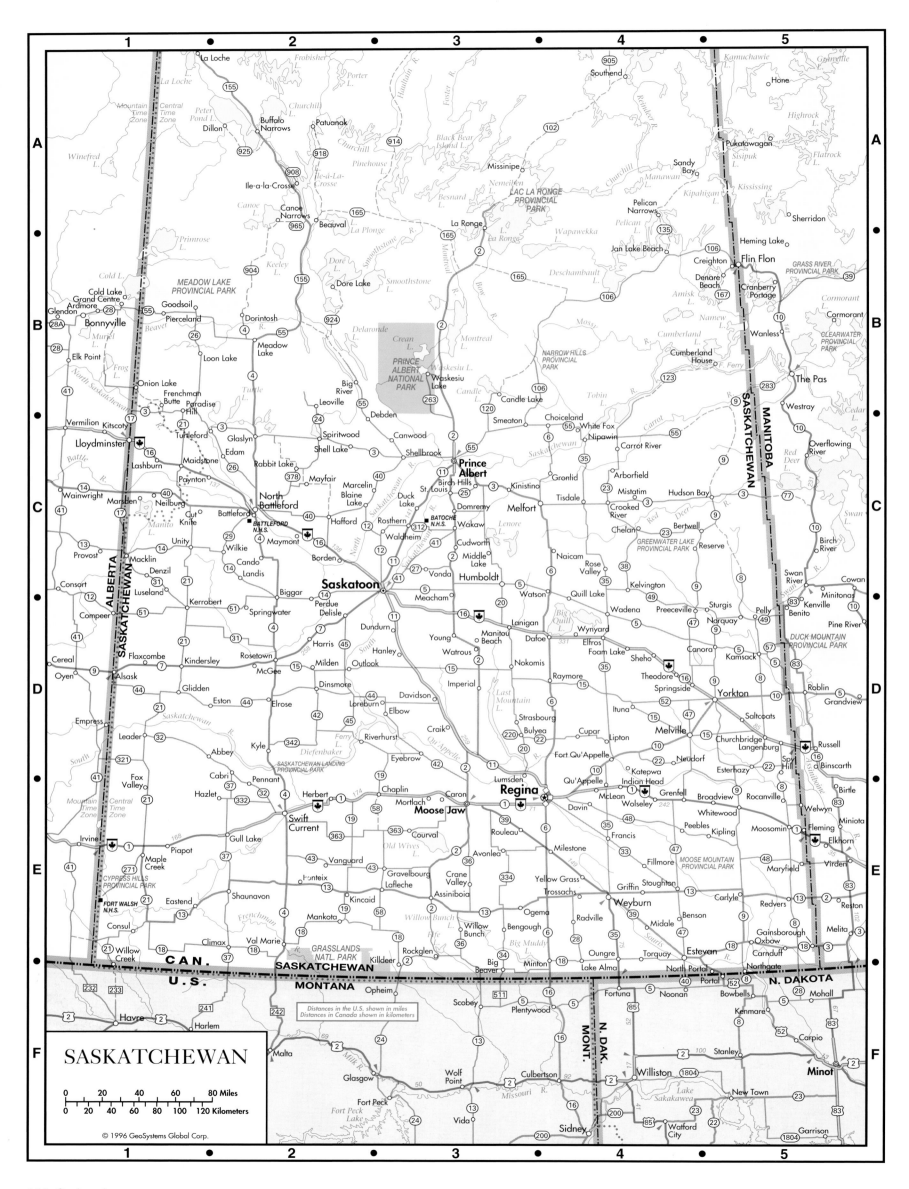

SASKATCHEWAN

0   20   40   60   80 Miles

0   20 40 60 80 100 120 Kilometers

© 1996 GeoSystems Global Corp.

Distances in the U.S. shown in miles
Distances in Canada shown in kilometers

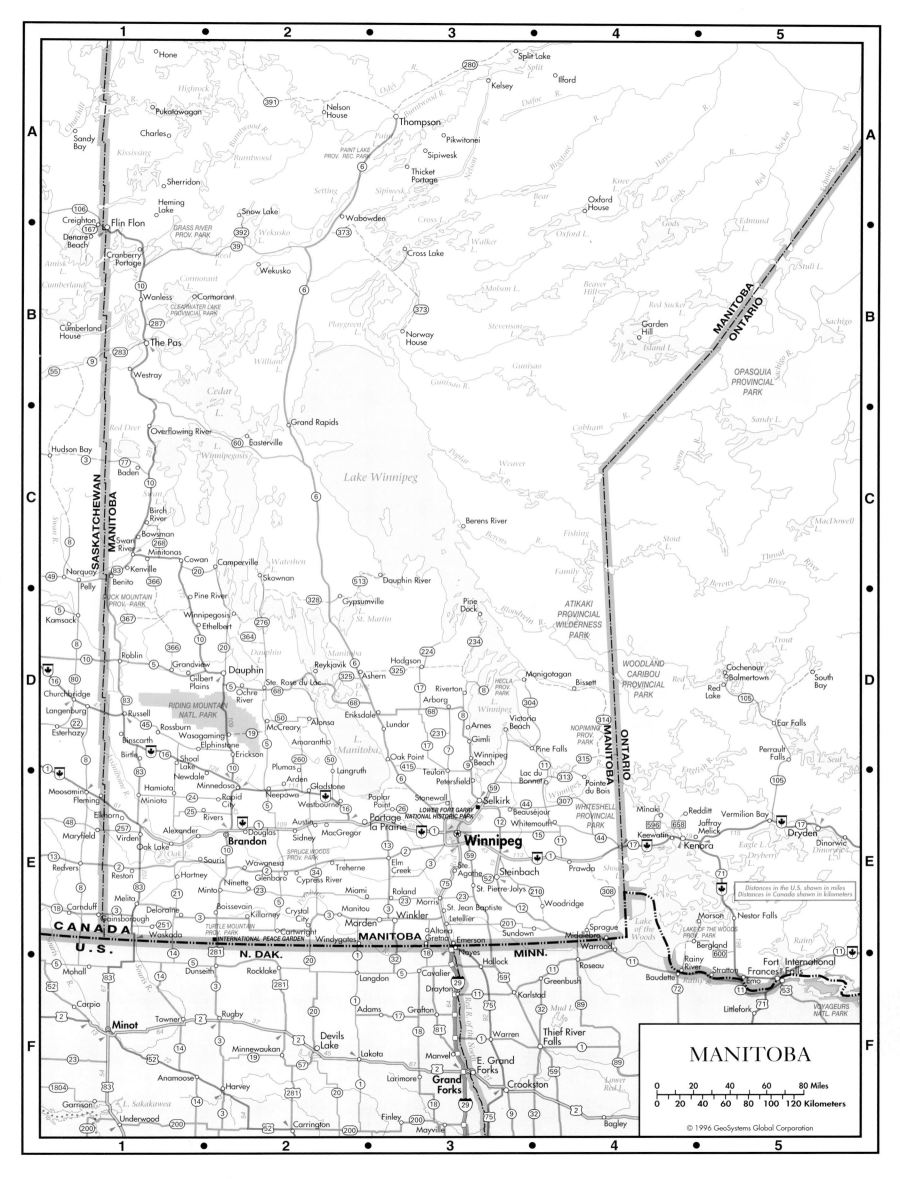

MANITOBA

Distances in the U.S. shown in miles
Distances in Canada shown in kilometers

| 0 | 20 | 40 | 60 | 80 Miles |

| 0 | 20 | 40 | 60 | 80 | 100 | 120 Kilometers |

© 1996 GeoSystems Global Corporation

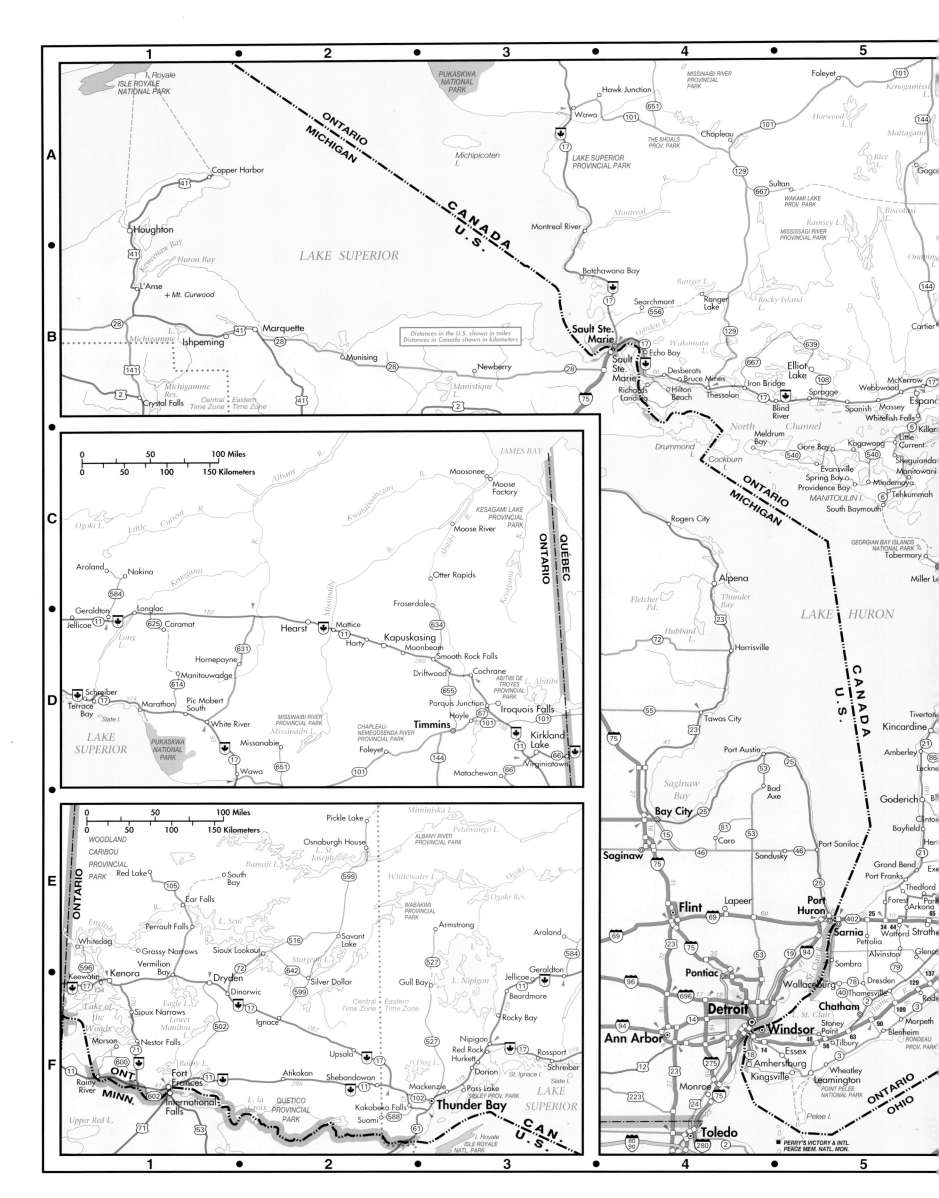

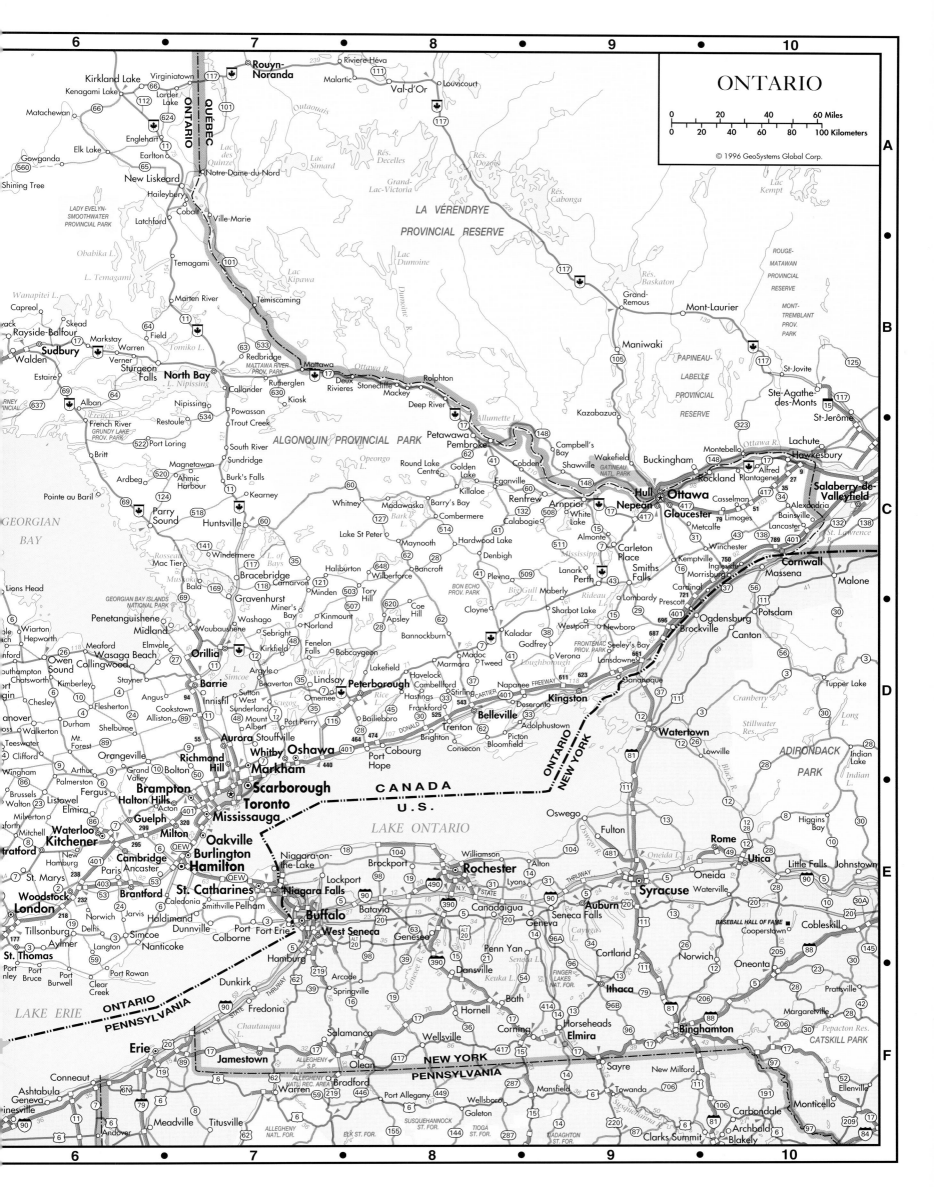

ONTARIO

0   20   40   60 Miles
0 20 40 60 80 100 Kilometers
© 1996 GeoSystems Global Corp.

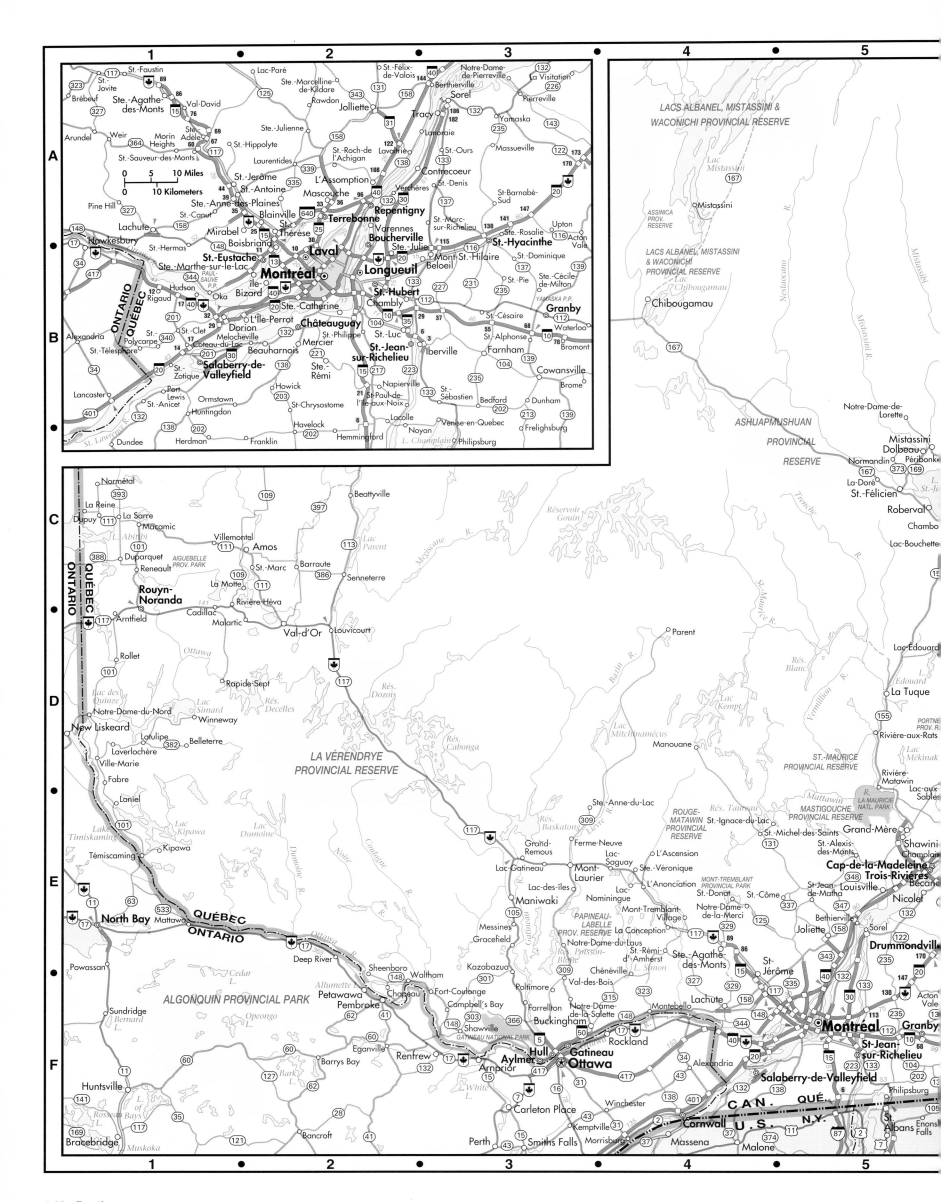

160 Québec

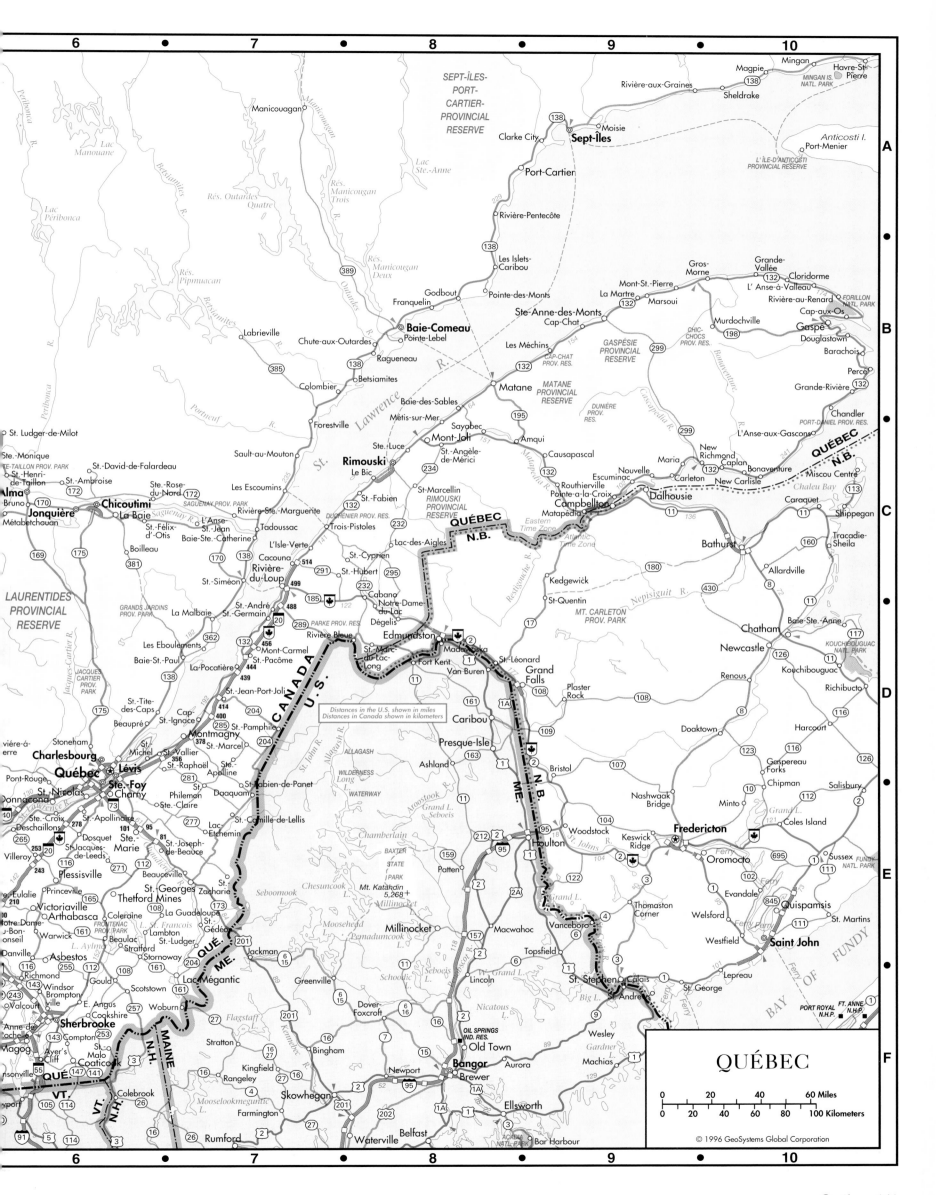

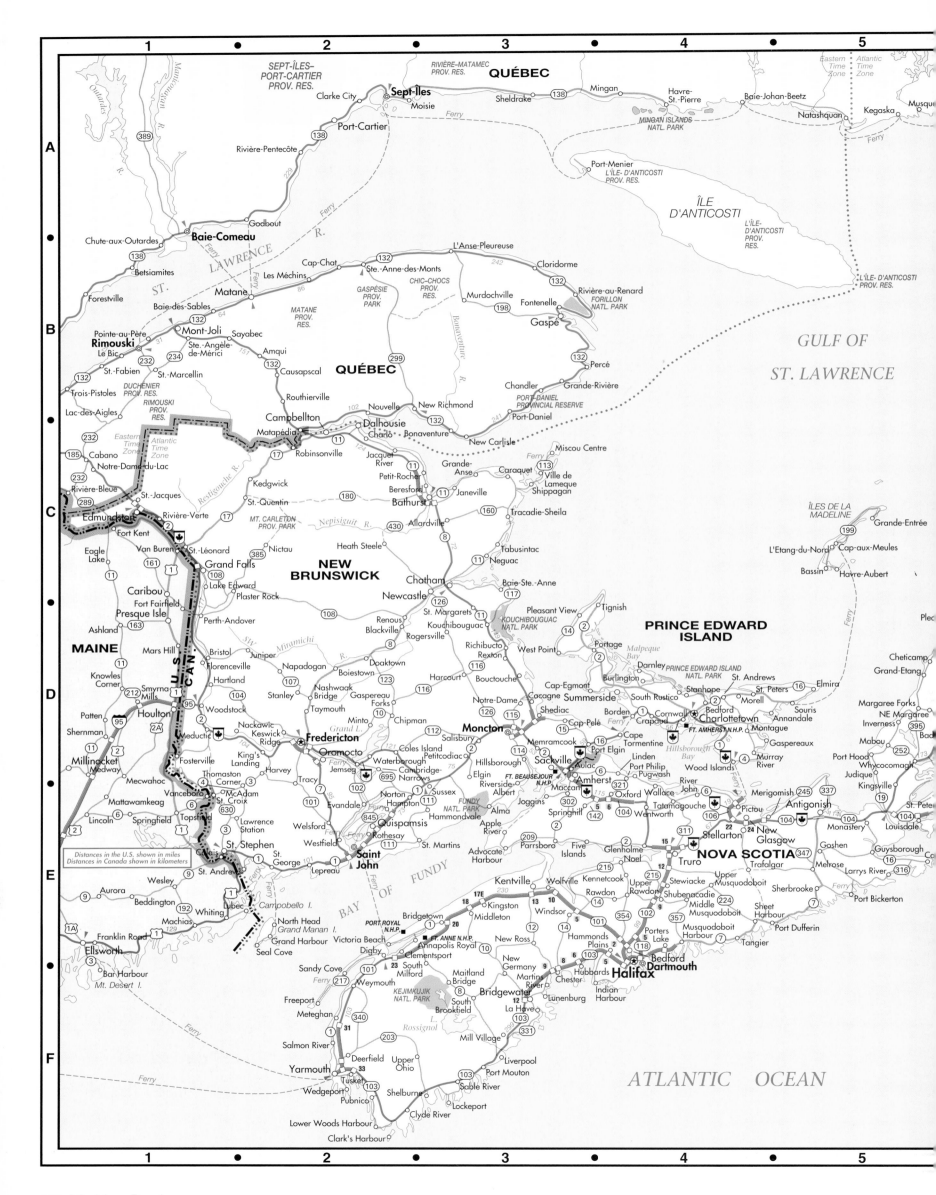

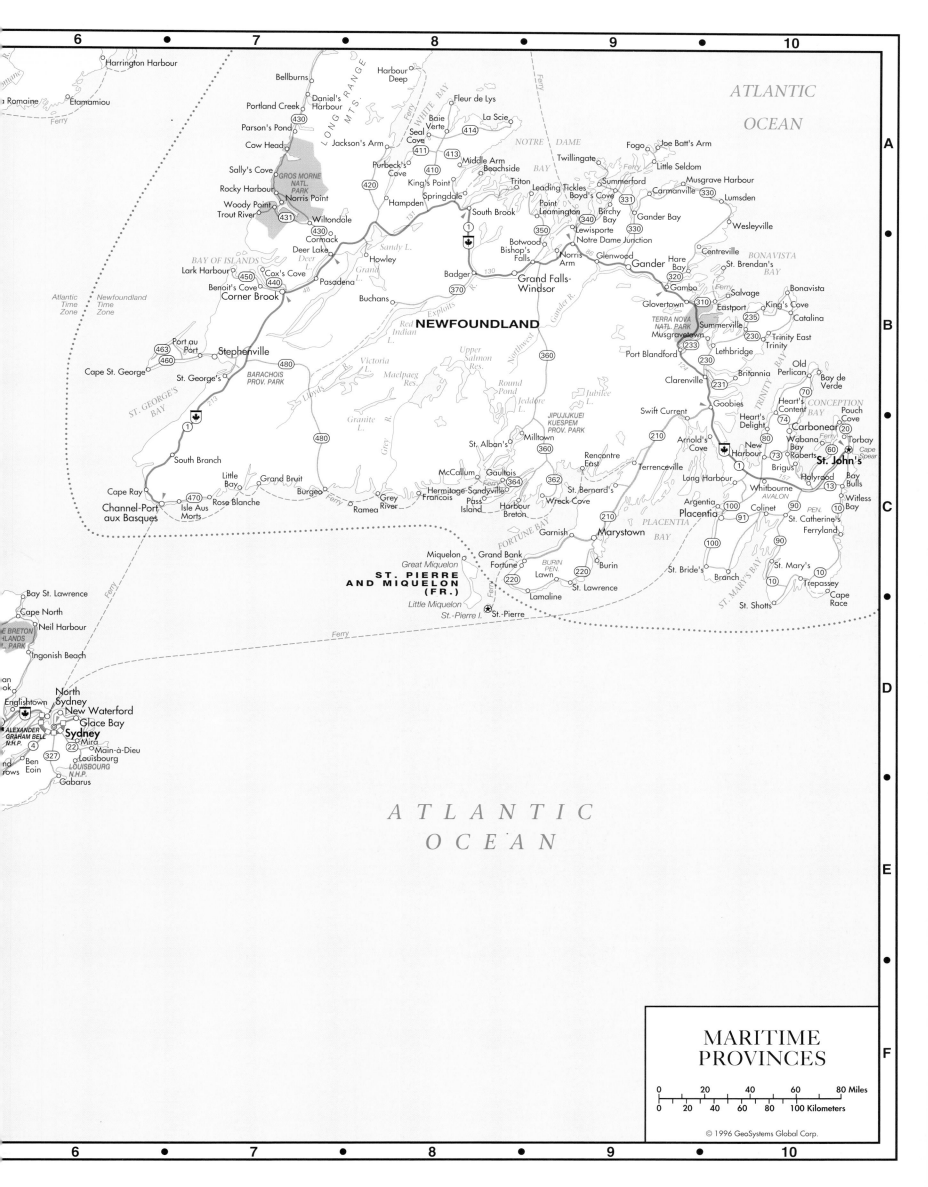

ATLANTIC OCEAN

NEWFOUNDLAND

Harrington Harbour
Etamamiou
a Romaine

Bellburns
Daniel's Harbour
Portland Creek
Parson's Pond 430
Cow Head
Sally's Cove
Rocky Harbour
GROS MORNE NATL. PARK
Norris Point
Woody Point
Trout River 431
Wiltondale
Cormack 430
Deer Lake
BAY OF ISLANDS
Lark Harbour 450
Cox's Cove 440
Benoit's Cove
Corner Brook
Pasadena
Howley
Grand L.
Deer L.
Buchans
Red Indian L.
Victoria L.
Maelpaeg Res.
Lloyds R.
Grey R.
Granite L.
Jeddore L.
Jubilee L.
Round Pond
Upper Salmon Res.

Harbour Deep
Fleur de Lys
La Scie
Seal Cove 411
Baie Verte 414
Purbeck's Cove
Middle Arm
Beachside
Jackson's Arm
King's Point 410
Hampden
Springdale
South Brook 1
Triton
Leading Tickles
Boyd's Cove
Point Leamington 340
Lewisporte 350
Notre Dame Junction
Bishop's Falls
Botwood
Norris Arm
Grand Falls-Windsor 370
Badger 130
WHITE BAY
NOTRE DAME BAY
Twillingate
Summerford
Birchy Bay 331
Gander Bay 330
Glenwood
Gander 320
Hare Bay
Gambo 310
Fogo
Joe Batt's Arm
Little Seldom
Musgrave Harbour
Carmanville
Lumsden
Wesleyville
Centreville
St. Brendan's
BONAVISTA BAY
Bonavista
Salvage 235
Eastport
King's Cove
Catalina
Trinity East
Trinity
Glovertown
Summerville
TERRA NOVA NATL. PARK
Musgravetown 233
Port Blandford 230
Lethbridge 230
Clarenville 231
Britannia
Old Perlican
Bay de Verde
Goobies
CONCEPTION BAY 70
Heart's Content 74
Heart's Delight
Pouch Cove
Carbonear 80
Wabana 20
Swift Current 210
Arnold's Cove
New Harbour 73
Brigus
Argentia 100
Placentia
Colinet 91
Whitbourne 90
St. Catherine's
Ferryland
Branch 10
St. Mary's
St. Shotts
Cape Race
Trepassey 10
Witless Bay
Bay Bulls 13
Holyrood
Torbay 60
St. John's
Cape Spear
AVALON PEN.
PLACENTIA BAY
Long Harbour
Terrenceville
Rencontre East
Wreck Cove
St. Bernard's
Garnish
Marystown 210
Burin 220
Lawn
St. Lawrence
Lamaline
FORTUNE BAY
Grand Bank
Fortune 220
BURIN PEN.
Milltown 360
St. Alban's
McCallum
Gaultois
Hermitage
François
Sandyville 364
Pass Island
Harbour Breton
JIPUJIJKUEI KUESPEM PROV. PARK
Channel-Port aux Basques
Isle Aux Morts 470
Rose Blanche
Burgeo
Grey River
Ramea
Grand Bruit
Little Bay
South Branch
Cape Ray
Stephenville 480
St. George's
Cape St. George
Port au Port 463 460
BARACHOIS PROV. PARK
ST. GEORGE'S BAY
Atlantic Time Zone
Newfoundland Time Zone

MIQUELON
Great Miquelon
ST. PIERRE AND MIQUELON (FR.)
Little Miquelon
St.-Pierre I.
St.-Pierre

Bay St. Lawrence
Cape North
Neil Harbour
Ingonish Beach
CAPE BRETON HIGHLANDS NATL. PARK
North Sydney
New Waterford
Glace Bay
Sydney
Mira
Main-à-Dieu
Louisbourg
LOUISBOURG NATL. PARK
Gabarus
Ben Eoin 327
Englishtown
ALEXANDER GRAHAM BELL N.H.P.
rows 4

ATLANTIC OCEAN

MARITIME PROVINCES

0 20 40 60 80 Miles
0 20 40 60 80 100 Kilometers

© 1996 GeoSystems Global Corp.

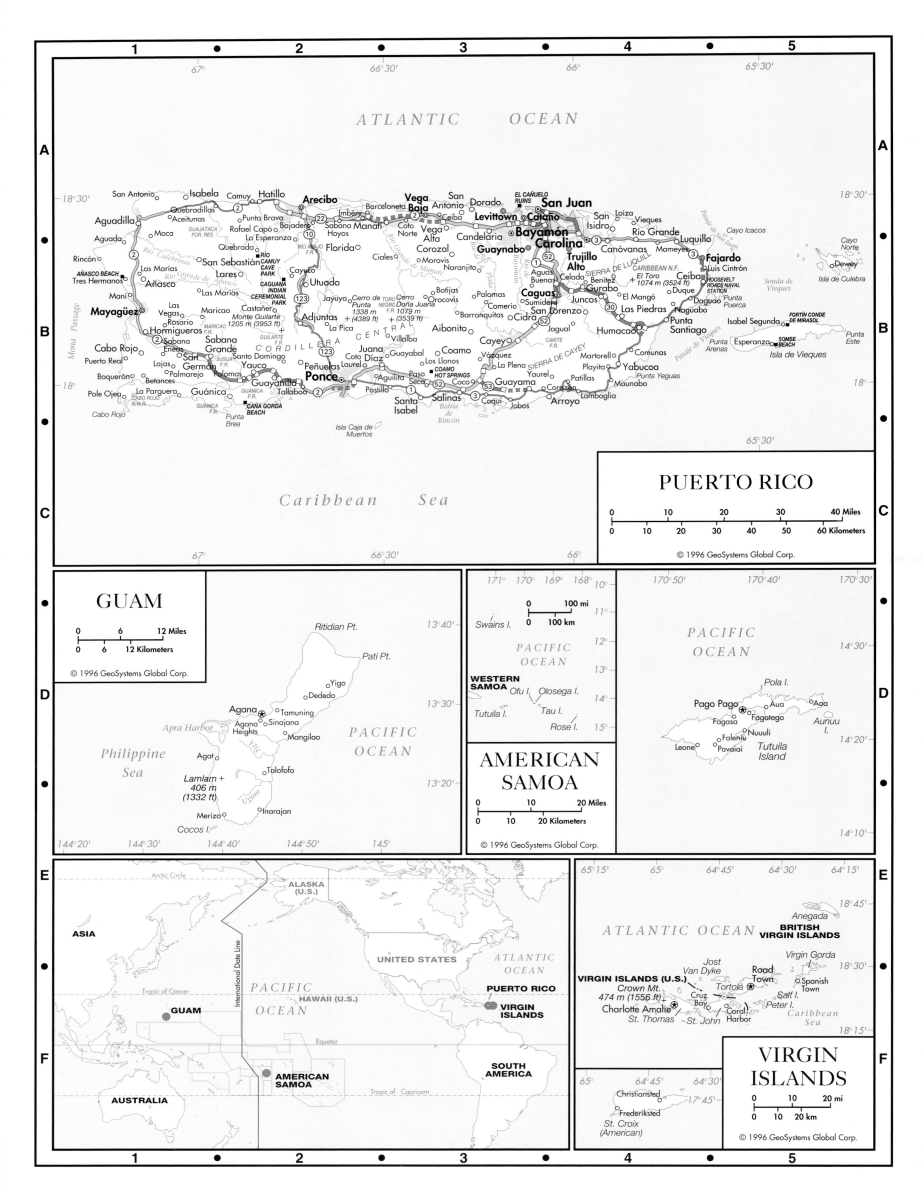

PUERTO RICO

0 10 20 30 40 Miles
0 10 20 30 40 50 60 Kilometers
© 1996 GeoSystems Global Corp.

GUAM

0 6 12 Miles
0 6 12 Kilometers
© 1996 GeoSystems Global Corp.

AMERICAN SAMOA

0 10 20 Miles
0 10 20 Kilometers
© 1996 GeoSystems Global Corp.

VIRGIN ISLANDS

0 10 20 mi
0 10 20 km
© 1996 GeoSystems Global Corp.

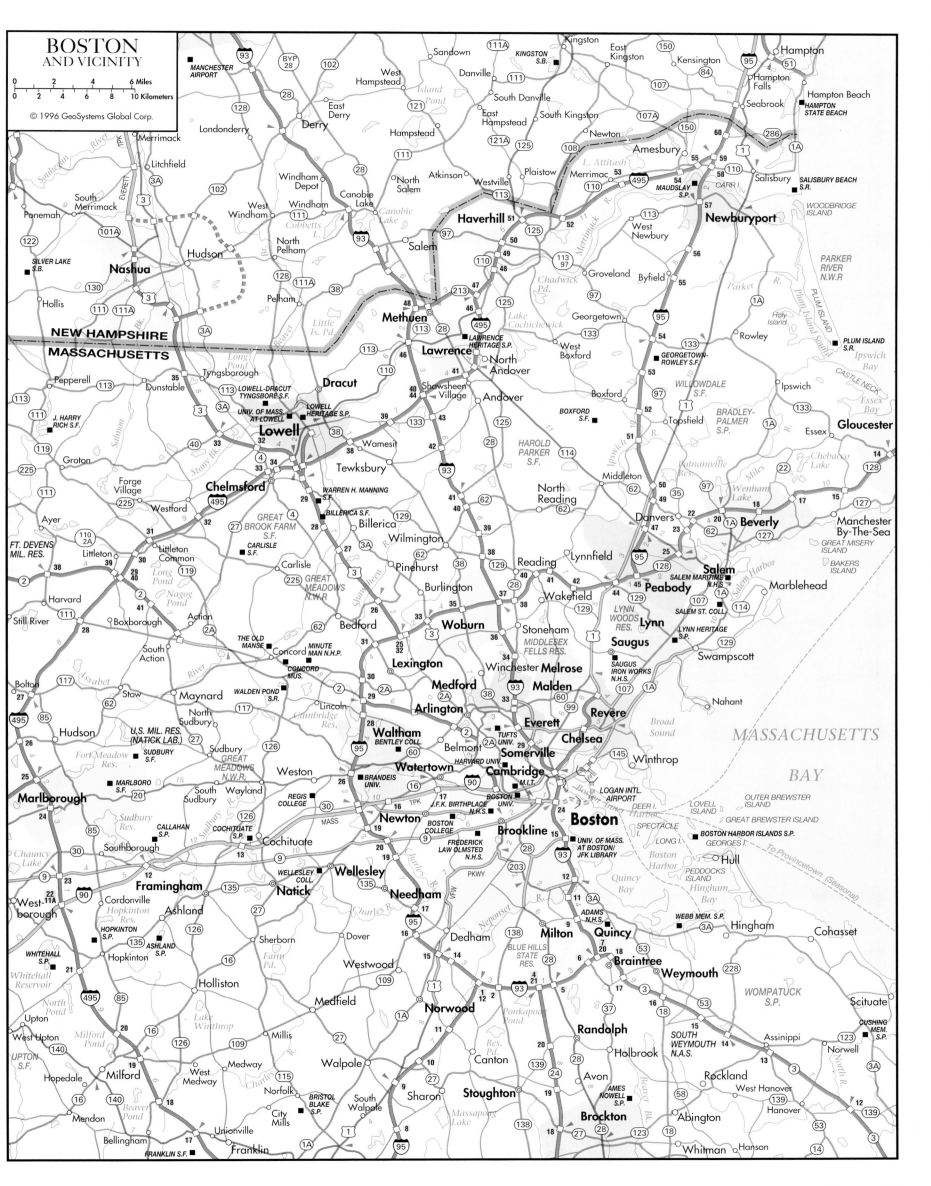

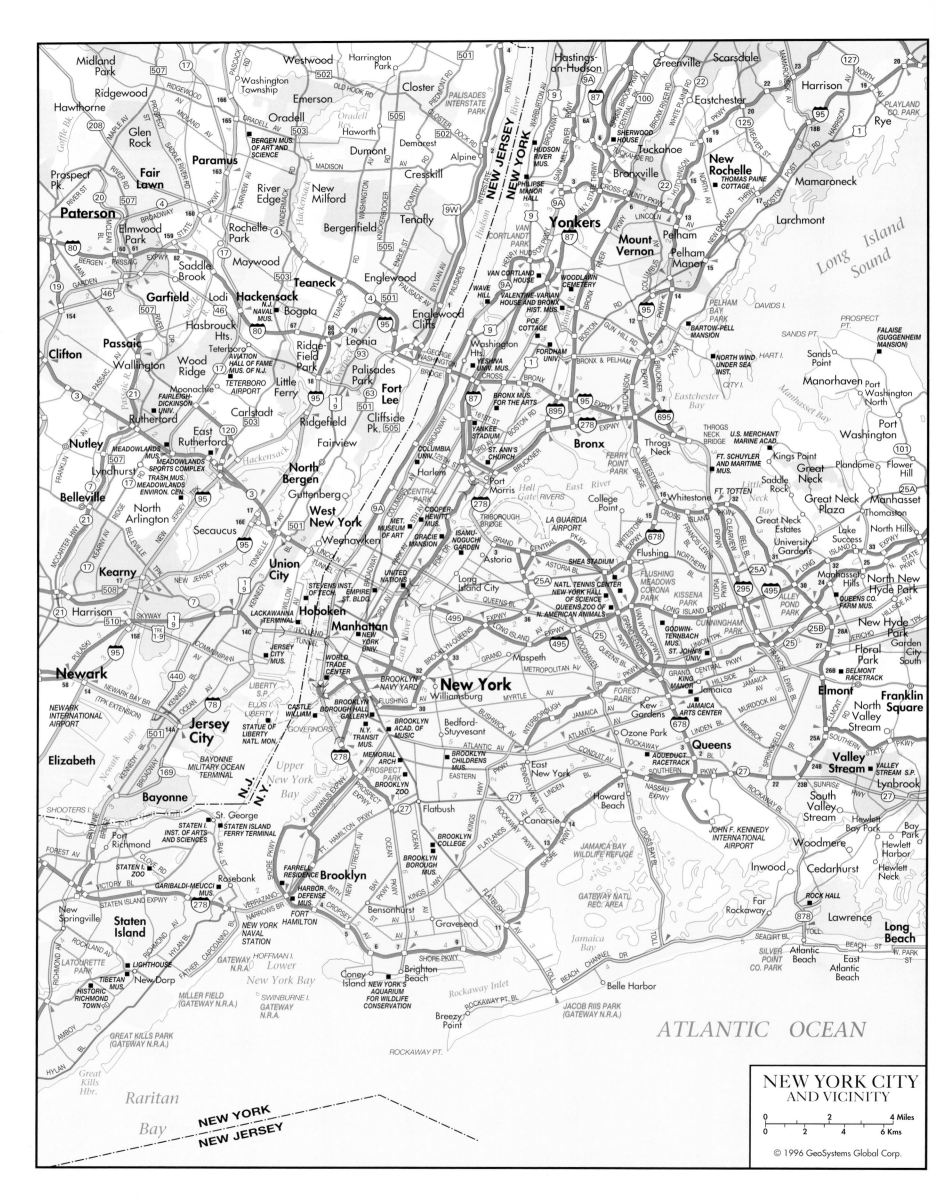

NEW YORK CITY
AND VICINITY

0   2   4 Miles
0   2   4   6 Kms

© 1996 GeoSystems Global Corp.

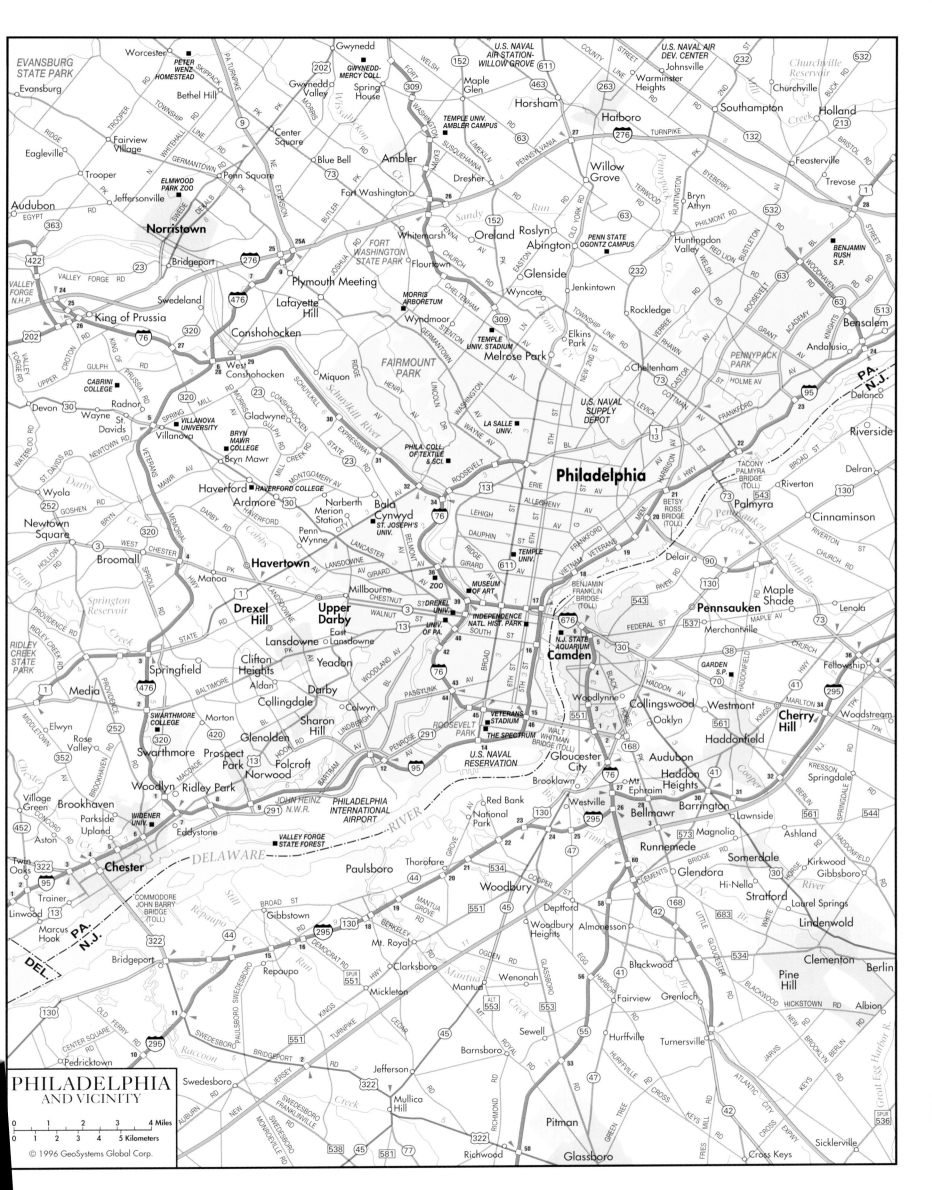

PHILADELPHIA
AND VICINITY

0   1   2   3   4 Miles
0 1 2 3 4 5 Kilometers

© 1996 GeoSystems Global Corp.

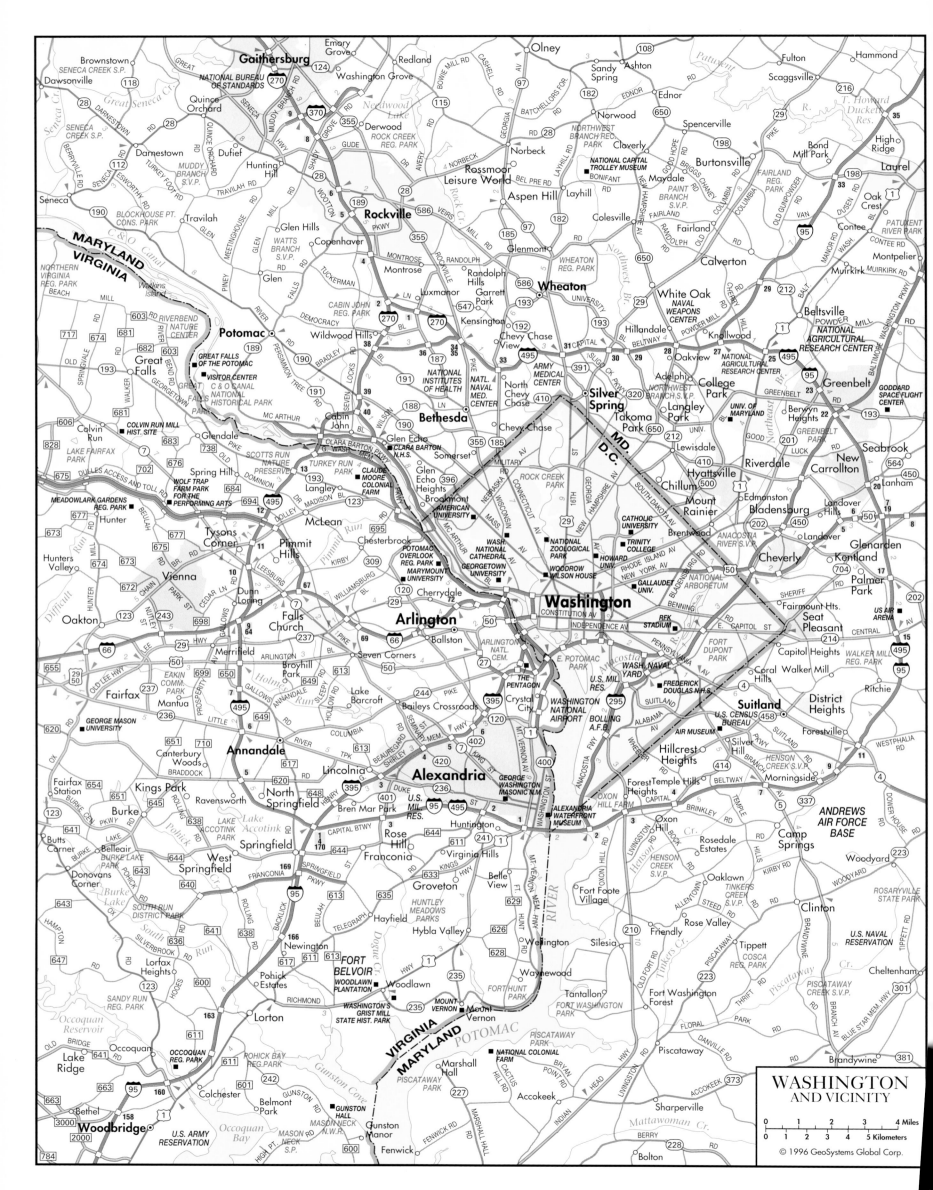

WASHINGTON
AND VICINITY

0   1   2   3   4 Miles
0   1   2   3   4   5 Kilometers

© 1996 GeoSystems Global Corp.

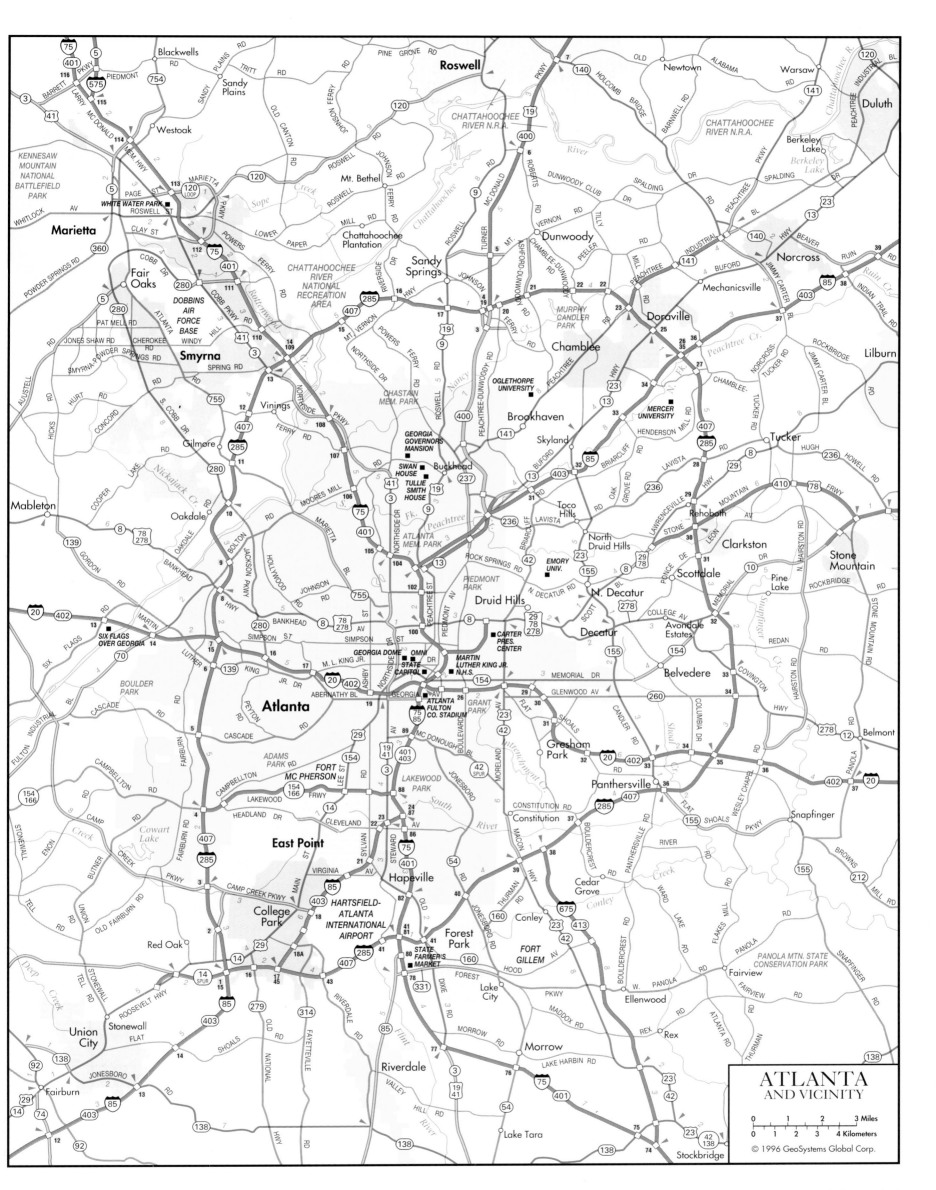

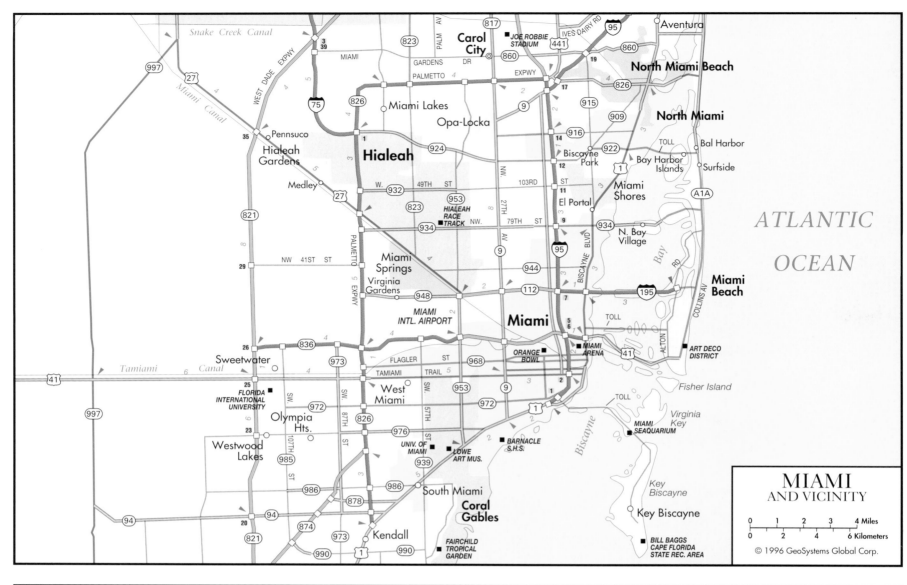

MIAMI
AND VICINITY

0 1 2 3 4 Miles
0 2 4 6 Kilometers

© 1996 GeoSystems Global Corp.

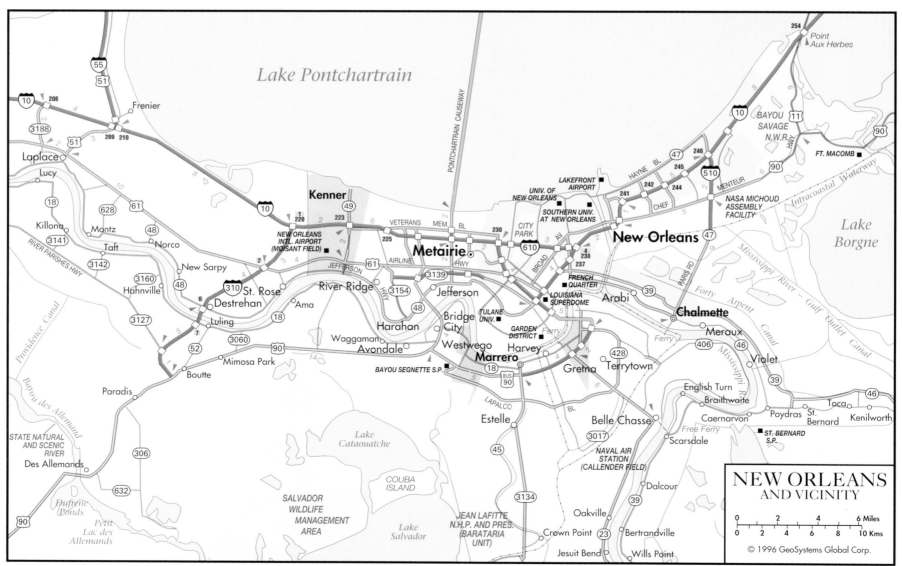

NEW ORLEANS
AND VICINITY

0 2 4 6 Miles
0 2 4 6 8 10 Kms

© 1996 GeoSystems Global Corp.

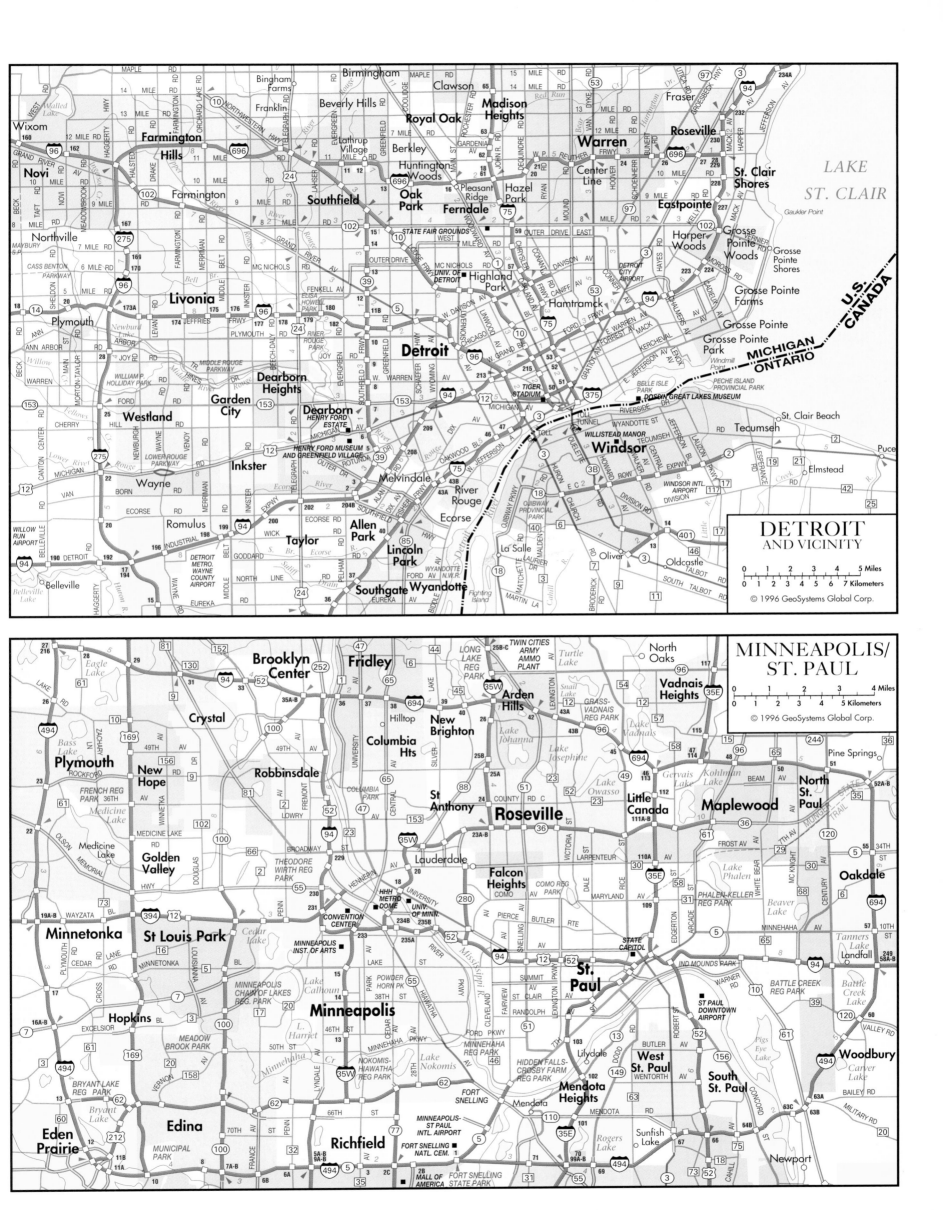

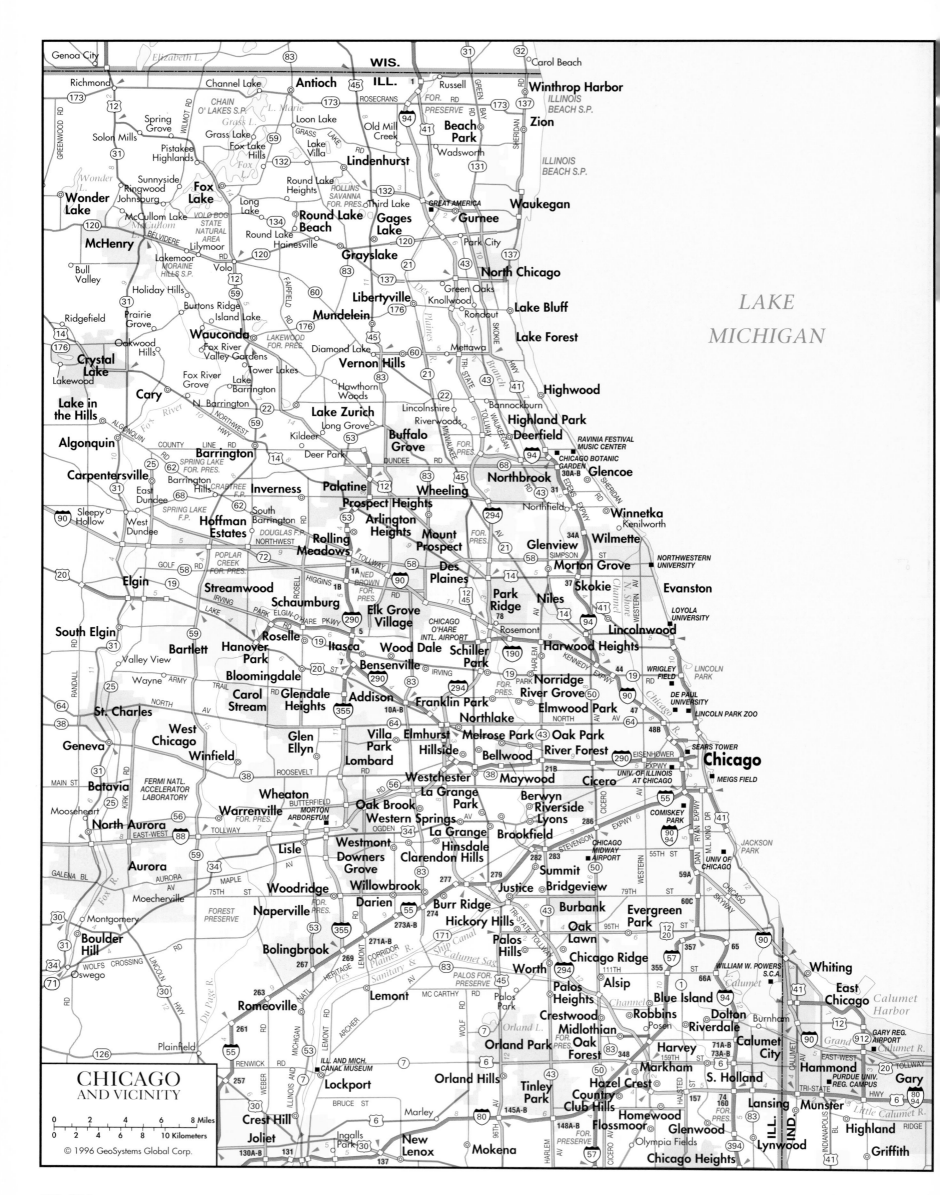

CHICAGO
AND VICINITY

0 2 4 6 8 Miles
0 2 4 6 8 10 Kilometers

© 1996 GeoSystems Global Corp.

LAKE MICHIGAN

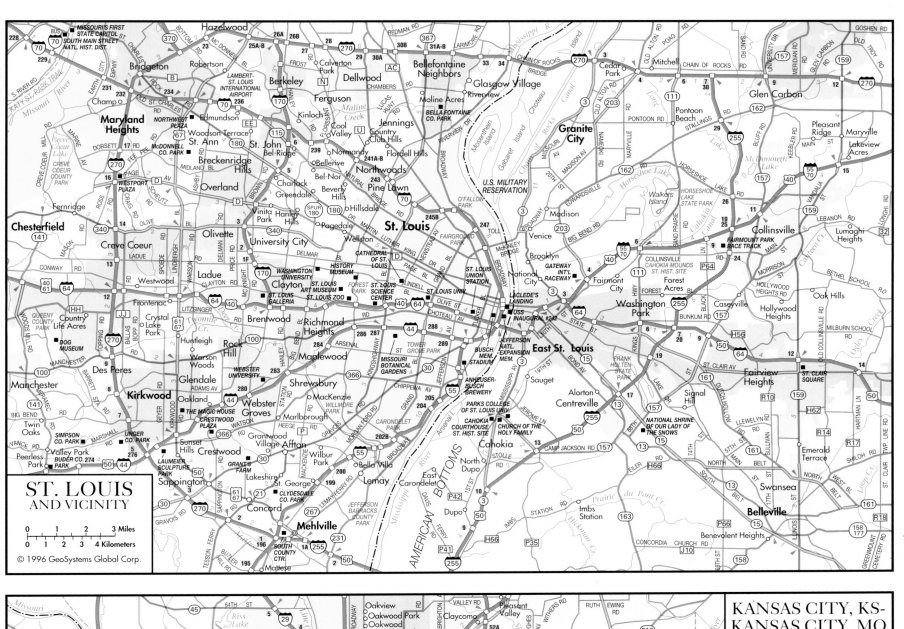

ST. LOUIS
AND VICINITY

0   1   2   3 Miles
0   1   2   3   4 Kilometers
© 1996 GeoSystems Global Corp.

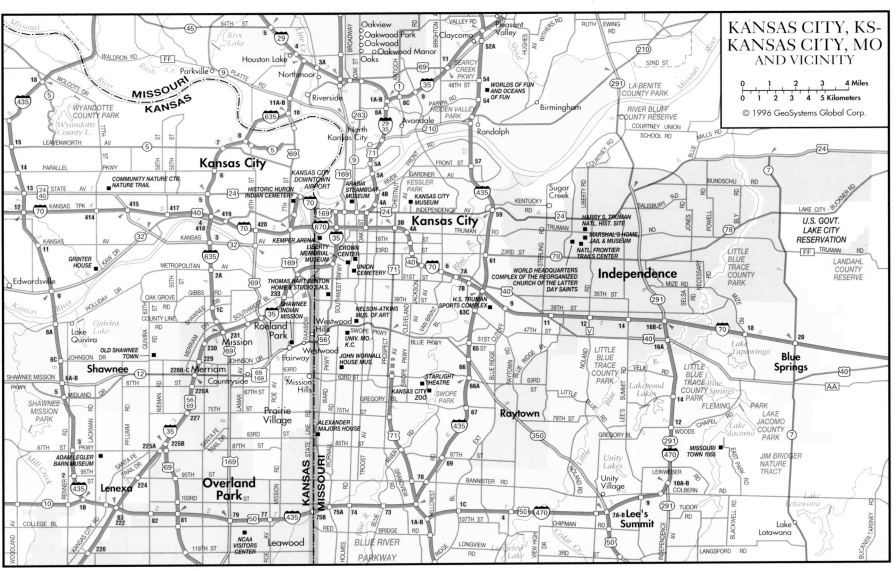

KANSAS CITY, KS-
KANSAS CITY, MO
AND VICINITY

0   1   2   3   4 Miles
0   1   2   3   4   5 Kilometers
© 1996 GeoSystems Global Corp.

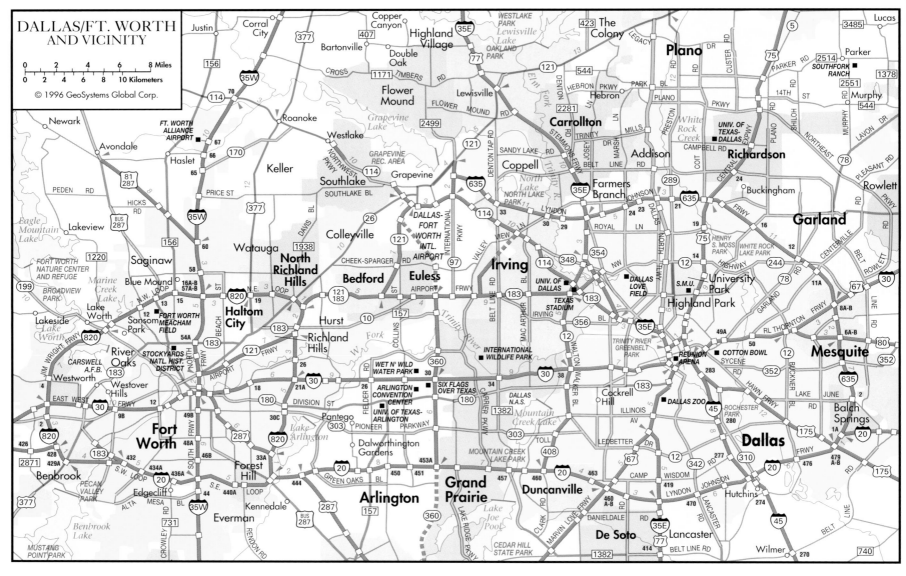

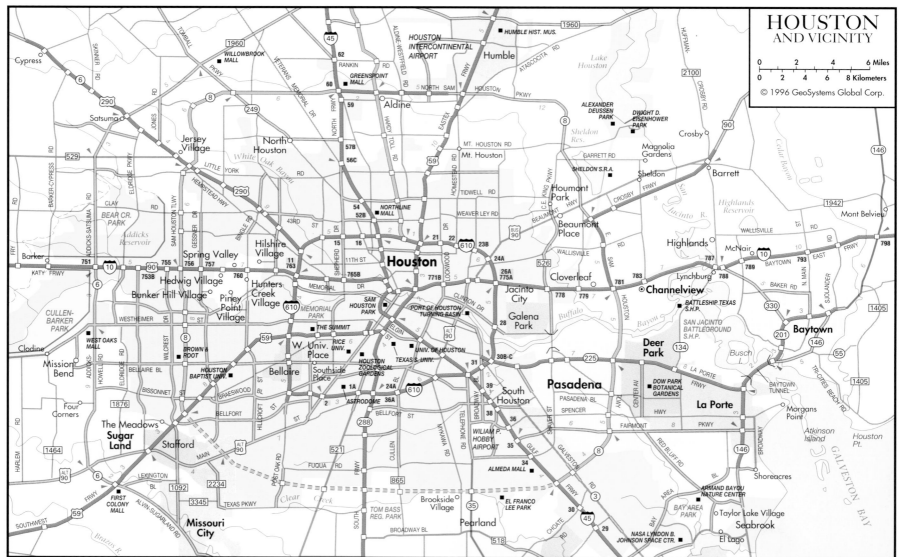

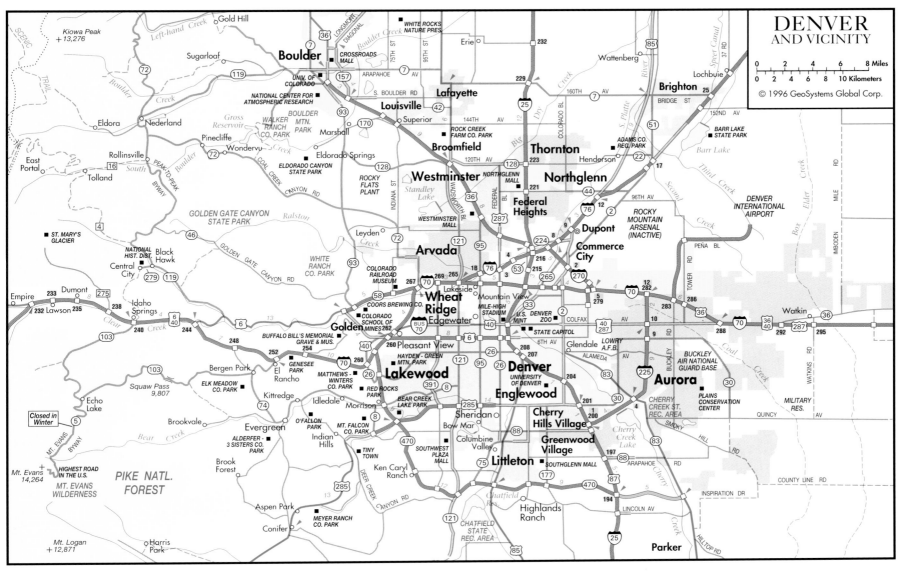

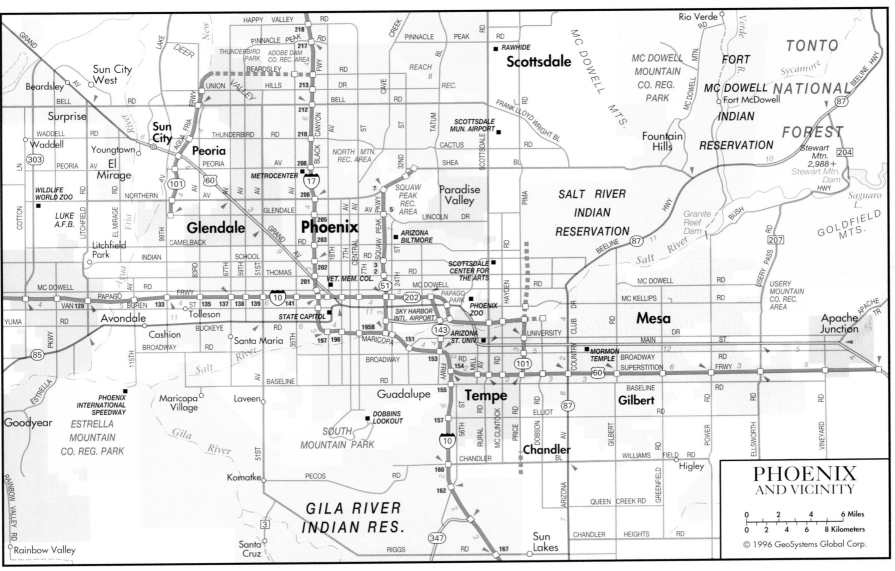

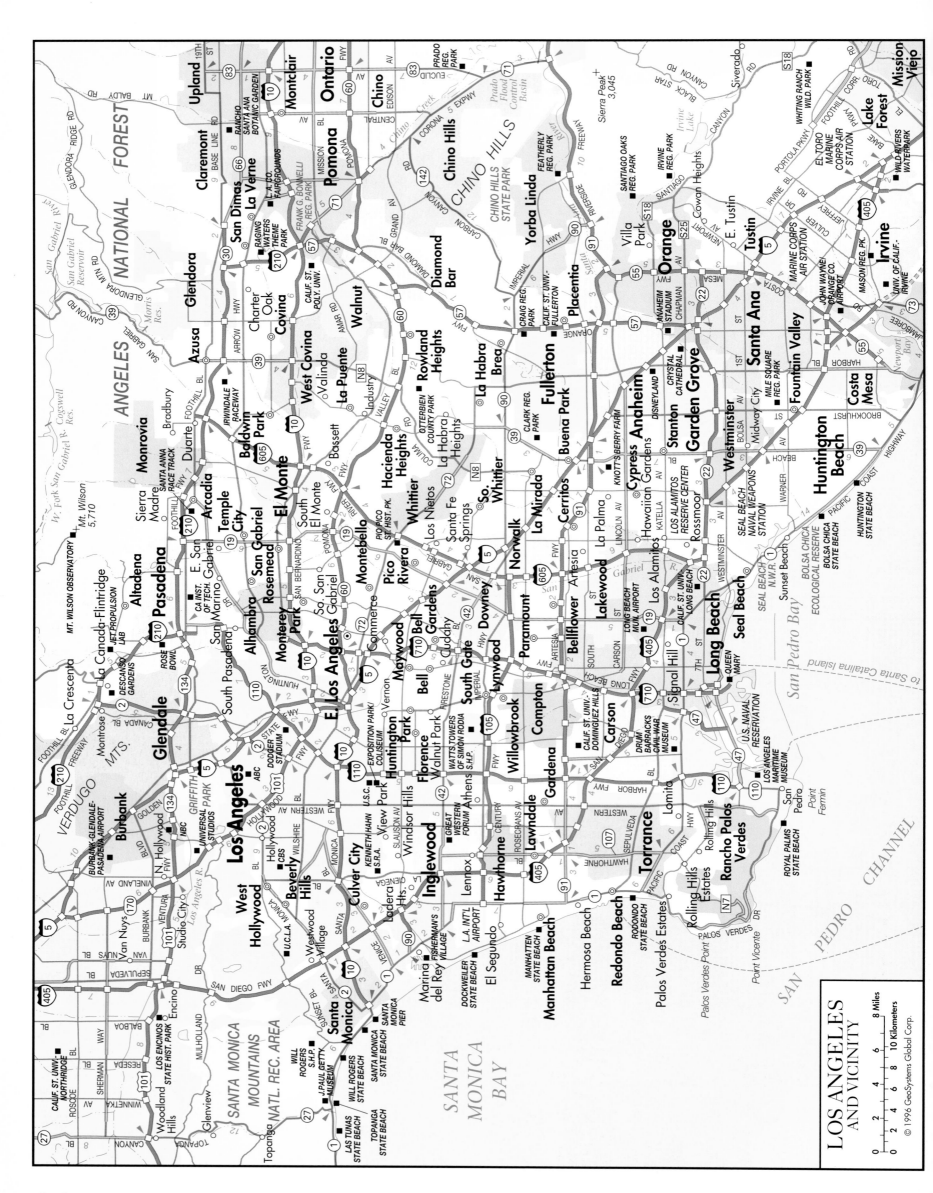

LOS ANGELES
AND VICINITY

© 1995 GeoSystems Global Corp.

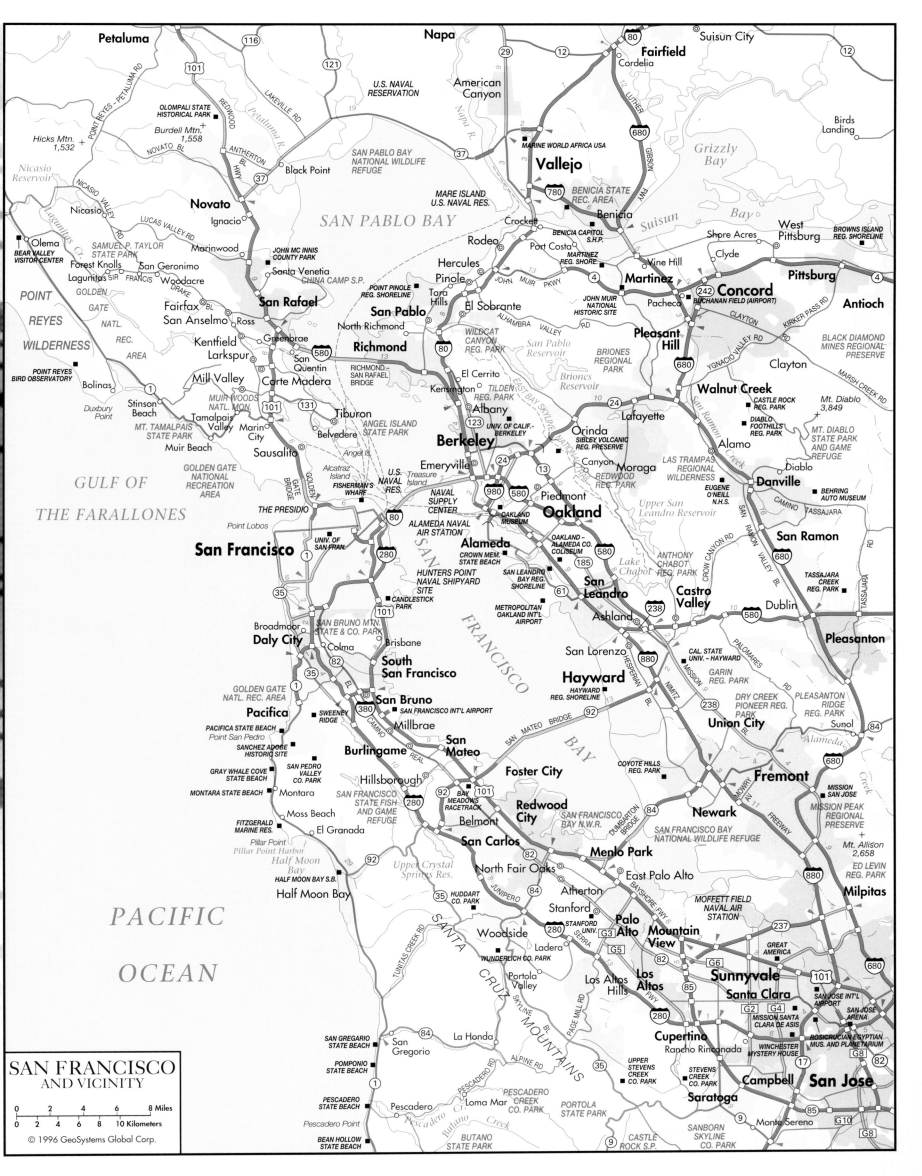

SAN FRANCISCO
AND VICINITY

0   2   4   6   8 Miles
0 2 4 6 8 10 Kilometers

© 1996 GeoSystems Global Corp.

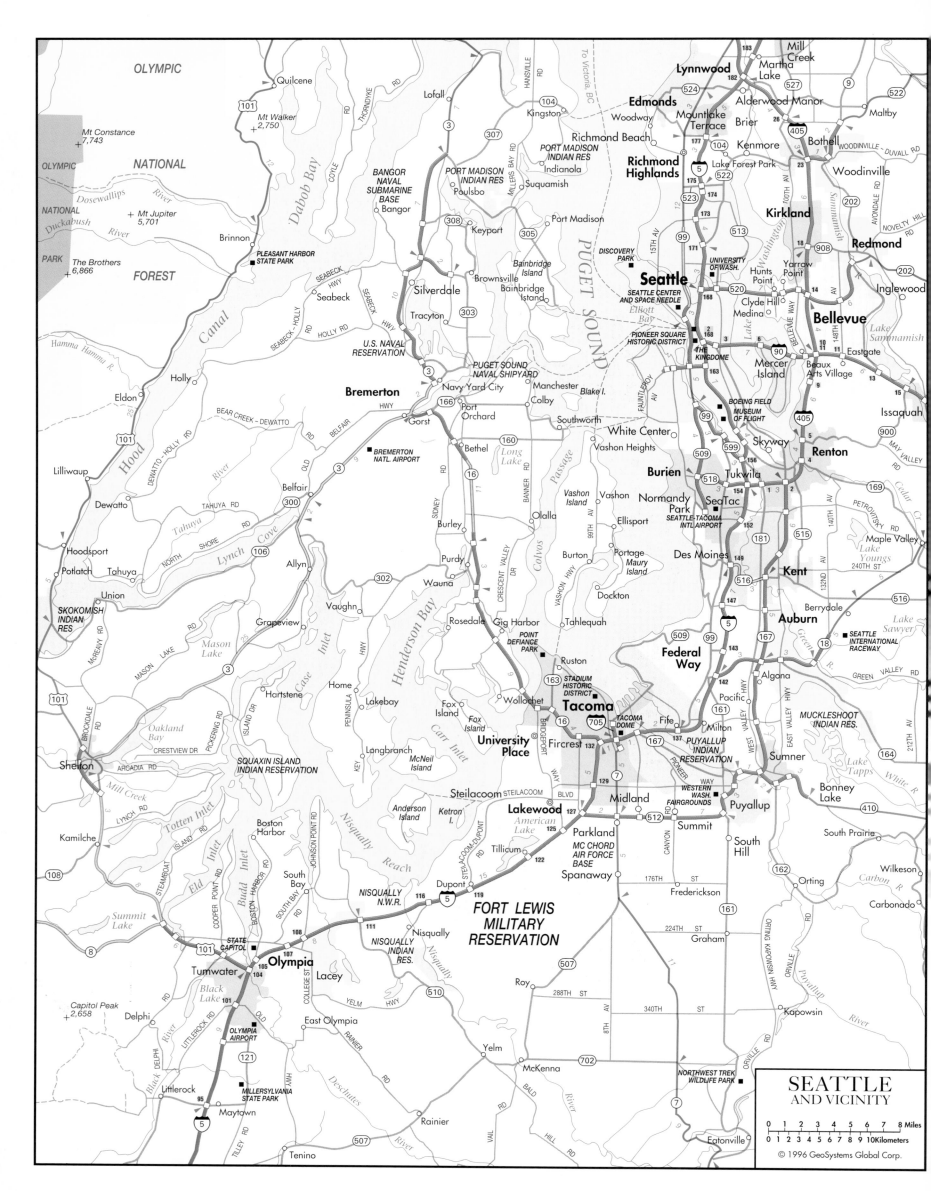

181

188

MINNESOTA
Page 123

MISSISSIPPI
Page 126

WISCONSIN
Page 150

WYOMING
Page 151

MANITOBA
Page 157

MARITIME PROVINCES
(New Brunswick, Newfoundland
Nova Scotia, Prince Edward Island)
Pages 162–163

206